Optical Interferometry - A Multidisciplinary Technique in Science and Engineering

Edited by Mithun Bhowmick

Published in London, United Kingdom

Optical Interferometry - A Multidisciplinary Technique in Science and Engineering
http://dx.doi.org/10.5772/intechopen.97946
Edited by Mithun Bhowmick

Contributors
Marco Chiarini, Alberto Parini, Gian Giuseppe Bentini, Carlos Frajuca, Rahim Ullah, Raja Yasir Mehmood Khan, Muhammad Faisal, Michael Saville, Rogério Dionísio, Mithun Bhowmick

First published in London, United Kingdom, 2022 by IntechOpen
IntechOpen is the global imprint of INTECHOPEN LIMITED, registered in England and Wales, registration number: 11086078, 5 Princes Gate Court, London, SW7 2QJ, United Kingdom

British Library Cataloguing-in-Publication Data
A catalogue record for this book is available from the British Library

Additional hard and PDF copies can be obtained from orders@intechopen.com

Optical Interferometry - A Multidisciplinary Technique in Science and Engineering
Edited by Mithun Bhowmick
p. cm.
Print ISBN 978-1-80355-702-1
Online ISBN 978-1-80355-703-8
eBook (PDF) ISBN 978-1-80355-704-5

Meet the editor

Dr. Mithun Bhowmick completed his Ph.D. in physics at Virginia Tech before becoming a faculty member in the Physics Department at Nazarbayev University (Kazakhstan) and a post-doctoral research associate at the University of Illinois, Urbana-Champaign (USA). He is currently a member of the faculty at the Miami University, Middletown (USA) campus. Dr. Bhowmick's research group focuses on spectroscopy in shock-compressed materials where specimens of a wide variety are experimentally studied under extreme conditions. He is interested in applications of short- and long-term pressure-temperature effects in the matter and associated sciences at the nanoscale.

Contents

Preface

This book reviews recent developments in optical interferometry (OI) techniques and aids readers in their exploration of different aspects of this subject. The focus of the novel reports presented in this volume ranges from traditional topics to newer applications, such as those related to biology and clinical procedures. Each chapter includes an introduction providing the reader with relevant background material.

Chapter 1 serves as a brief introduction to the subject, highlighting the rapidly expanding nature of the technique. Chapter 2 introduces the concept of integrated optics and photonics in sensing applications, with specific reference to IO devices. Chapter 3 focuses on the engineering aspect of OI techniques and discusses a novel 2D phase reconstruction method for the measurement of plasma properties where the signal-to-noise ratio is unstable. Chapter 4 highlights fiber-optic applications in tunable filters and sensors using a fiber Bragg grating. Application areas are extended to optically active materials in Chapter 5, where the authors note recent developments and the importance of polarization-based OI techniques applied to food components. Chapter 6 presents a general overview of interferometric gravitational wave detectors.

I would like to thank all the contributors to this book for sharing their knowledge and expertise with us. I am very grateful to Ms. Sara Debeuc of IntechOpen for all her help and support. I hope this book helps readers to realize the powerful and versatile techniques and applications related to OI.

Mithun Bhowmick
Mathematical and Physical Sciences,
Miami University,
Middletown, USA

Chapter 1

Introductory Chapter: Optical Interferometry in Interdisciplinary Applications

Mithun Bhowmick

1. Introduction

Optical interferometry (OI) involves the superposition of electromagnetic waves to form interference fringes where tiny changes in optical path difference cause significant variation in the intensity pattern of the detected light. The precision and flexibility associated with modern-day interferometric techniques have enabled researchers to apply them to a wide range of applications for solving problems related to science and engineering [1–7]. In the recent years, applications have become more sophisticated. For example, optical interferometric methods are now useful for detecting early chemistry, to analyze 3D surface topography, in biosensing, and in several topics related to investigate the matter under extreme conditions [6–10]. The use of OI in astronomy should be mentioned separately due to the volume of research that has been reported over the last few decades [11]. In this chapter, a brief overview on interferometric applications will be presented. The focus will be on the most recent advances from several fronts of science and engineering to note the width of applications where OI techniques have been relevant. The examples are by no means exhaustive due to the limited scope of this chapter. The mentions here serve only pointers for future discussions or an extended review on the state-of-the-art of OI techniques.

In most simple terms, in OI, a coherent light beam is split into two using a beam splitter. The two beams cover different trajectories, commonly known as "paths" in optics before superposing to create interference fringes. The fringes or intensity patterns depend on phase differences between the two beams, resulting from the different lengths of the paths taken by them. Classification of interferometers could be done in many ways, depending on the focus and purpose. If the carrier frequency is of importance, two major groups named homodyne and heterodyne could be mentioned based on if the frequency has changed in the process.

Likewise, interferometers could be classified into major groups. Most OI devices that are being discussed here are double-path interferometers, where the signal and reference beams travel along distinct paths before recombining. A common path interferometer is, as their name suggests, example where the two beams travel the same paths.

2. Typical applications of OI in science and engineering

Interferometry resulting from electromagnetic waves has been used in many routine applications. One of the most popular methods in OI involves the superposition

IntechOpen

of Doppler-shifted laser beams. For simplicity and convenience of discussion, applications of interferometry could be divided into three major areas of science and engineering research. Historically physics and astronomy have used interferometry in many ways and could be taken as the first major application area. The second group of applications could be grouped under the engineering and applied science header. Finally, applications focused on research problems related to biology makes the third major area. Of course, this oversimplified classification is insensitive to some nuances of interdisciplinary works and serves only an ad hoc purpose.

These techniques found extensive relevance in investigations where probing small displacements, refractive index changes, surface irregularities, and remote analysis of optical signals are needed [11]. The progress of technology has played a critical role in shaping the advancement of these techniques. Even the "successful" demonstration of the famous negative result in physics from the 19th century, named after Michelson and Morley, found extraordinary level of precision with the help of modern technologies [12]. Interferometers have typically been used in the scientific industry for manufacturing and characterization of surfaces, integrated circuits, multiplexing methods needed for applications in telecommunications, in navigation systems, and to aid spectroscopy for materials characterization [13].

Extremely active and rapidly progressing, the typical applications provide the foundations for many interdisciplinary applications of OI techniques. These examples, noted in Section 3, are the most awe-inspiring results of modern-day scientific research.

3. Some selected interdisciplinary examples

In this section, some selected applications/techniques are highlighted as examples. Once again, this selection is not exhaustive, but representative of interesting emerging techniques.

Spatial and temporal resolutions are at the heart of modern-day research questions. A relatively recent technique based on the phase of fluorescence signals, named as fluorescence interferometry, has emerged over the last few years where meso and nanoscale resolutions were achieved in biological specimens [9]. The method is fully optical and simple in terms of instrumentation. Typically, the setup consists of excitation, lens pairs for scanning, and detection units. The self-referencing interferometer collects fluorescence from the emission from both sides of the specimen, when the specimen is being translated along the optical axis. An interference pattern is created when the coherence conditions are met from the depth of the sample, thus generating a spatial map. Considering the challenges of noninvasive clinical procedures, including but not limited to precision and cost of potential automation in medical fields, fluorescence interferometry could open up new horizons in imaging/sending applications.

High time and space resolution at real time in ambient conditions is challenging. Achieving the same under extreme conditions is understandably more challenging, since the detector of light gets destroyed in the process [7]. OI using a laser beam and Doppler-shifted light forms a branch known as "laser interferometry". Examples include popular and established methods such as VISAR, ORVIS, and stress gauges [6]. One of the novel applications of such systems is in shock compression studies, where finding velocities of free surface and embedded layers have been extremely important [7]. During a shock compression process, with a powerful shockwave progressing along the depth of a sample at ~ km/s velocities, it is critical to know these parameters inside the specimen. There are several aspects to it. First, as described above, the specimen cannot host the detector "device", since it would be destroyed after one single-stage shockwave experiment. Second, the geometry, opacity, and nature of shock response vary widely from one sample system to another, making it very difficult to design and implement the technique. A solution to those challenges was to keep the detecting device outside of the specimen, and to vary the laser wavelength according to the opacity of the material, as governed by the transmission spectra. In many materials (solids and liquids) this works due to their transparency at telecommunication wavelengths such as 1550 nm. Moreover, the already-established technology at those wavelengths makes it easy to overcome engineering challenges, especially related to signal-to-noise ratios. This technique, known as photon Doppler velocimetry (PDV), uses a single-mode laser beam, which gets split into two. One of them is then focused on the surface whose velocity is being measured. For example, in case of a projectile moving in free space, the beam would be focused on the nearest surface of it, incident normally. This beam can be reflected easily if the surface is polished or coated with metal films. The second beam from the beam splitter is the reference, with which the reflected beam is superposed for creating the interference pattern. By counting the beat frequency of the pattern once can generate a velocity history of the moving surface. The same technique can be applied to any surface inside the specimen if one is interested to know the evolution of shockwave amplitude. In this case, a thin layer (~ 30 nm) of the highly reflective surface is created at the surface from where the shockwave would be detected. By repeating the experiment with samples of different thicknesses, a complete picture of the shockwave propagation could be created. Shockwave propagation has many applications if the spatial and temporal resolution is high, for example, when it is approximately in the ns range. Early chemistry and photo-physics, mechano-optical manifestations of stress-related events, and impedance changes are only a few example applications [14–16].

Another notable application area of OI is interferometer-based sensors. Highly versatile, compact and sensitive, these sensors use to extract a wide variety of information such as temperature and refractive index changes in materials. In most common setups, Fabry-Perot systems, cascaded structures, parallel interferometric structures, fiber Bragg grating incorporated interferometer, and a Mach-Zehnder system with fiber-optic components could be found [17]. This class of interferometers is particularly effective in complementary studies to static and dynamic pressure experiments due to their sensitivity to strain-related effects in the matter. It is possible to extract information on temperature and strain (or effects) simultaneously from such measurements, and often more complex applications are possible with creatively hybridizing this technique with other related OI techniques (such as combining a Fabry-Perot with a Mach-Zehnder system).

Small sample detection, specifically related to biosensing, is another major area based on optical sensing applications where OI has contributed significantly. Traces of chemicals as small as micro or picolitres could be sensed using a technique where microfluidic devices are used in combination with OI [18–20]. This technique has enabled real-time monitoring of reaction systems and product formations, useful for novel biochemical applications in the microscopic scale. A related more application has used a ring cavity ultrafast laser to achieve small sample detection. This system employed a fiber optic Michelson interferometer in tandem with optofluidic devices to achieve label-free hybridization and sensing of DNAs with high precision [21].

The above examples are only representative applications where OI was used in overlapping areas of scientific research. A vast pool of examples could be easily found in literature where uses of OI in physics and astronomy, telecommunications, holography, navigation systems, atomic clocks, and other time/frequency domain applications could be found. Apart from the biosensing applications mentioned above, techniques such as phase contrast imaging based on differential interferometry have been developed.

4. Summary

OI has been a strong application tool in almost every discipline of science and engineering. As devices based fundamentally on the superposition of light and all-optical, noninvasive signal processing, these methods have remarkable flexibility and relevance. In this chapter, an overview of the most recent applications has been presented. While typical applications in science and engineering research have seen remarkable updates in the past few years, some novel applications have also emerged as a result of collaborative efforts in cross-disciplinary projects.

Acknowledgements

The author is thankful for the resources provided by the Miami University Regionals for conducting research.

Conflict of interest

The author declares no conflict of interest.

Notes/thanks/other declarations

None.

Author details

Mithun Bhowmick
Miami University, Middletown, Ohio, USA

*Address all correspondence to: bhowmim@miamioh.edu

References

[1] Bhowmick M, Ullrich B. Introductory Chapter: Interferometry. In: Interferometry - Recent Developments and Contemporary Applications. London, United Kingdom: IntechOpen; 2019 [Online]. Available: https://www.intechopen.com/chapters/65820 DOI: 10.5772/intechopen.84371

[2] Patel R, Achamfuo-Yeboah S, Light R, Clark M. Optics Express. 2014;**22**(22):27094-27101

[3] Bhowmick M, Nissen EJ, Dlott DD. Journal of Applied Physics. 2018;**124**:075901

[4] Chen LC, Nguyen XL. Measurement Science and Technology. 2010;**21**(5)

[5] Domachuk P, Grillet C, Ta'eed V, Mägi E, Bolger J, Eggleton BJ, et al. Microfluidic interferometer. Applied Physics Letters. 2005;**86**:024103

[6] Forbes JW. Shock Wave Compression of Condensed Matter: A Primer. Maryland, USA: Springer; 2012. DOI: 10.1007/978-3-642-32535-9

[7] Bhowmick M, Basset WP, Matveev S, Salvati L III, Dlott DD. Optical windows as high-speed shock wave detectors. AIP Advances. 2018;**8**:125123

[8] LASL. In: Marsh SP, editor. Shock Hugoniot Data. Berkeley and Los Angeles, California, USA: University of California Press; 1980

[9] Bilenca A, Cao J, Colice M, Ozcan A, Bouma B, Raftery L, et al. fluorescence interferometry: principles and applications in biology. Annals. New York Academy of Sciences. 2008;**1130**:68-77

[10] Kaiser F et al. Quantum enhancement of accuracy and precision in optical interferometry. Light: Science and Applications. 2018;**7**:17163

[11] Vorburger TV, Song J, Petraco N, Lilien R. Emerging technology in comparisons. Academic Press. 2019;**2019**:275-304

[12] Müller H, Herrmann S, Braxmaier C, Schiller S, Peters A. modern michelson-morley experiment using cryogenic optical resonators. Physical Review Letters. 2003;**91**:020401

[13] Hariharan. Basics of Interferometry. 2nd ed. Sydney, Australia: Elsevier Inc.; 2006. p. 9780080465456

[14] Nissen EJ, Bhowmick M, Dlott DD. Shock-induced kinetics and cellular structures of liquid nitromethane detonation. Combustion and Flame. 2021;**225**:5-12

[15] Nissen EJ, Bhowmick M, Dlott DD. Ethylenediamine catalyzes nitromethane shock-to-detonation in two distinct ways. The Journal of Physical Chemistry B. 2021;**125**(29):8185-8192

[16] Stekovic S, Springer HK, Bhowmick M, Dlott DD, Stewart DS. Laser-driven flyer plate impact: Computational studies guided by experiments. Journal of Applied Physics. 2021;**129**(19):195901

[17] Miura D, Asano R. Reference Module in Biomedical Sciences, Chapter: Biosensors: Immunosensors. Elsevier Inc.; 2021. DOI: 10.1016/B978-0-12-822548-6.00008-X

[18] Janik M, Sosnowska M, Gabler T, Koba M, My'sliwiec A, Kutwin M, et al. Life in an optical fiber: Monitoring of cell cultures with microcavity in-line

Mach-Zehnder interferometer. Biosensors and Bioelectronics. 2022;**217**:114718

[19] Baijin S, Qi B, Zhang F, Zhong L, Xu O, Qin Y. Hybrid fiber interferometer sensor for simultaneous measurement of strain and temperature with refractive index insensitivity. Optics Communications. 2022;**522**:128637

[20] Krześniak A, Gabler T, Janik M, Koba M, Jönsson-Niedziółka M, Śmietana M. A microfluidic system for analysis of electrochemical processing using a highly sensitive optical fiber microcavity. Optics and Lasers in Engineering. 2022;**158**:107173

[21] Hu X-g, Zhao Y, Peng Y, Tong R-j, Zheng H-k, Zhao a J, et al. In-fiber optofluidic michelson interferometer for detecting small volume and low concentration chemicals with a fiber ring cavity laser. Sensors and Actuators: B. Chemical. 2022;**370**:132467

Chapter 2

Integrated Optics and Photonics for Optical Interferometric Sensing

Marco Chiarini, Alberto Parini and Gian Giuseppe Bentini

Abstract

Integrated optics (IO) microsystems, based on guided wave and photonics structures, can find interesting developments for optical interferometric analysis in sensing applications. In fact, IO interferometric microsystems can act as transducers of the information carried on by an optical signal and originated along the signal optical path. In addition, the application of Integrated Optics and Photonics techniques, allow producing very small size and reduced power consumption instrumentation. These features explain the reason why the IO microsystems for interferometric analytics know an increasing interest in many areas such as Astrophysics, Environment, Biosciences, Space and Earth Exploration, Safety and Security. This chapter starts by synthetically describing the basics of the main analytical techniques covered by IO micro-devices. A discussion of the integrated micro-device fabrication procedures, with an analysis of the implied performance limitations, follows. Finally, a description of new generation optical interferometric microsystems, applied to different scientific and technical areas, completes the chapter.

Keywords: optical interferometry, Fourier transform interferometry, integrated optics, photonics, integrated optical sensors, Mach–Zehnder micro-interferometers, MEOS sensing devices

1. Introduction

There is a growing interest in finding detection technologies enabling real-time and on-line monitoring of many elements having remarkable, (in some cases dramatic), impacts on our everyday lives. Consequently, a huge effort has been devoted to designing and building instruments with capabilities that would not even be thought of a few years ago. Nowadays are required instruments to analyse many fields of high interest ranging from environmental pollutants or pathogens in air and water, up to the Security and Safety in the urban ambient or in the logistic chain of food production and transport, as well as many other applications.

In most cases, the realisation of this kind of device involves analytical techniques that should avoid chemical reactants, difficult to be used outside specialised laboratories. Moreover, the device has to be routinely applied by almost untrained people that should also be able to read and understand the measurement result. At the same time, the analytical equipment must be self-contained and available at a very low cost.

Nowadays, the only available solution suitable to satisfy the whole of these requirements is to develop highly miniature integrated equipment to allow mass production using technologies likewise the planar technologies driven by the development of microelectronics. At the present state of the art, most of these requirements can be satisfied when it is possible to develop sensing techniques belonging to the family of the optical integrated microdevices.

Optical detection can concern different properties of light such as Intensity, (in the case of photometric analyses), Wavelength, (in the case of spectroscopic analyses), Refractive Index, (in the case of Index change due to the presence of the analyte molecules at the Sensor Surface), etc.

A large variety of optical and photonic microdevices, based on interferometric, energy dispersive elements, photonic micro and nanofluidic devices, has been designed, studied and reported in the literature [1–6].

The development at industrial quality level of active optical materials, (particularly Lithium Niobate, $LiNbO_3$), joined to the use of planar technology, has allowed the fabrication of monolithic integrated optical microdevices having sensing capabilities comparable to the correspondent standard laboratory equipment.

As highlighted in the next paragraphs, IO and Photonics microfabrication techniques, offer the possibility to generate in monolithic and miniaturised systems, nearly all the equivalents of classical optical components (mirrors, splitters, combiners, phase shifters, etc.). Furthermore, such IO components can be integrated into and interfaced with guided wave circuits, such as integrated optical waveguides and optical fibres. These characteristics often strongly simplify all issues related to optical alignment and maintenance, the flow of the luminous signal, carrying the information, being firmly confined within fixed and well defined optical channels. In the following an overview of the characteristics and potentialities of IO devices is presented, in particular for a wide range of interferometric sensing applications.

2. Basic principles of IO for interferometric sensing

The basic element of any Integrated Optic device is the optical waveguide that can be generated by tailoring the refractive index (n) in the near-surface region of the base material. It must be remembered that the light transmission is confined in the regions where the refractive index is higher than in the surroundings.

Several techniques have been developed to obtain local variation of the refractive index creating the possibility to obtain an integrated optical waveguide. In this chapter, we will only mention some of the most common techniques used for $LiNbO_3$ substrate, which is one of the most used materials in optical device fabrication. One possibility is to use local doping processes obtained by photolithographic definition of the desired waveguide geometry associated with the dopant thermal diffusion (typically, Ti diffusion or Proton exchange processes), to increase the refractive index in the doped region. Alternatively, it is possible to create a waveguide by lowering the refractive index of the base material in the regions outside the waveguide introducing lattice damage through ion bombardment. A third possibility is to pattern the surface with the desired geometry, and etching the surrounding region to obtain a ridge waveguide protruding from the surface. **Figure 1** reports a sketch of the two geometries.

In general, the geometries of the integrated microsystems are designed to reproduce the same physical effects obtained by optical instrumental architectures created in laboratory, assembling several optical elements like mirrors, beam splitters, etc.

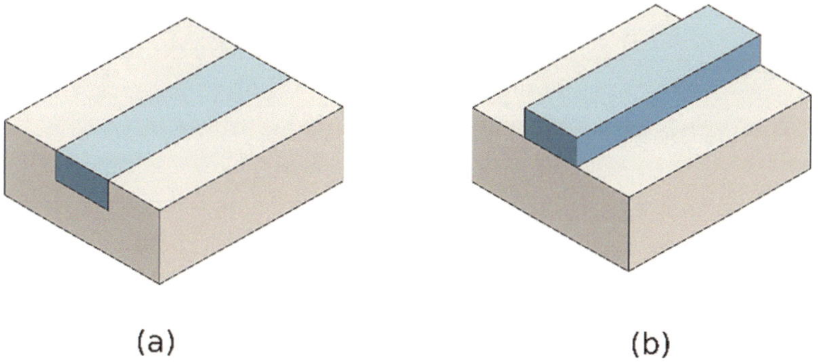

Figure 1.
Two basic geometries of integrated optics waveguides: (a) buried waveguide, (b) ridge waveguide.

Probably, the most frequently used integrated micro-spectrometric devices are based on the Mach-Zehnder Interferometric, (MZI), geometry [1–4] or on Young Interferometer, (YI) or, more recently, on the Staircase Micro Diffractive Gratings, (MDG) first developed by Michelson [5].

All these devices take advantage of the Electro-Optic properties of the $LiNbO_3$ substrate. In fact, the optoelectronic properties of the substrate allow to locally controlling the refractive index of a waveguide by applying a suitable electric field, so creating a Pockels cell that induces a phase modulation in the light transmitted in the specific waveguide (See **Figure 2**).

In particular, it can be useful to recall that, in the case of spectroscopic analysis techniques, the wavelength-dispersive systems, such as prisms or gratings, spatially spread the light wavelengths at different angles allowing the direct measurements of the relative intensities, (wavelength spectrum), by using suitable photodetectors at the corresponding angles.

On the contrary, in the Fourier Transform Spectroscopy the intensity of the total light beam that contains the whole ensemble of wavelength, at the same time, is measured. In this case, the measurement with a traditional Mach-Zehnder instrument is performed by splitting the light beam into two branches that are then recombined giving rise to an interference pattern. The light intensity of the recombined beams is

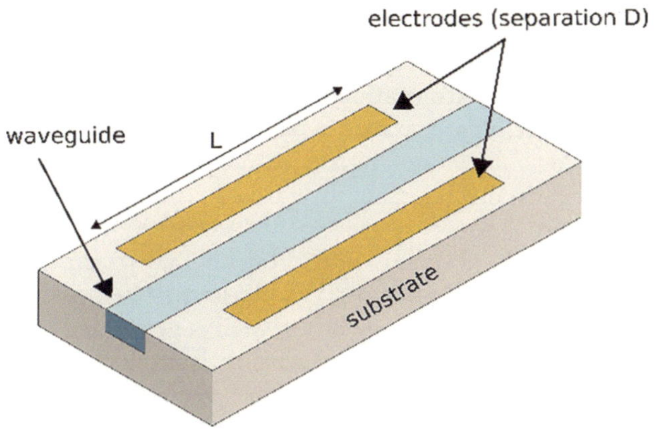

Figure 2.
Integrated version of a Pockels cell to be used as a phase modulator.

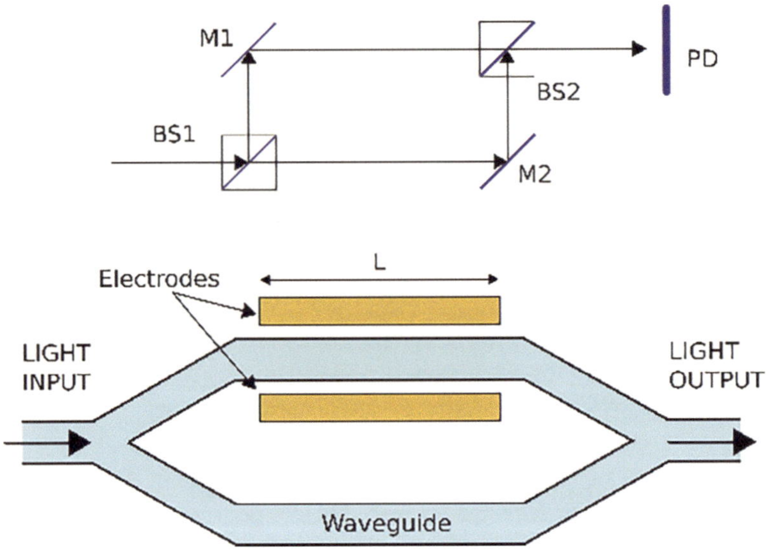

Figure 3.
Upper: Traditional Mach-Zehnder interferometer geometry. The light beam first crosses the BS1 beam splitter then, through the M1 and M2 mirrors, the two light beams are recombined in BS2 and then arrive in the photodetector PD. lower: Equivalent integrated optic device: The voltage applied to the electrodes controls the phase shift in the light beams propagating in the two arms.

measured as a function of the phase shift, generated by the respective different optical path-lengths $l = nL$, where n is the refractive index and L is the geometrical path length.

Figure 3(a), reports a typical example of Mach-Zehnder equipment, whereas **Figure 3(b)** is shown the equivalent device fabricated with planar technology. Following the optical interference laws, the intensity measured by the photodetector (PD) depends on the phase shift between the two optical paths. Inducing a variation in the optical paths, l, of one of the two arms the phase shift changes and, in turn, the correspondent variation of the light intensity is measured by the PD.

3. Photonic integrated interferometers for chemical and biological sensing

Actually, due to the relatively simple design, the Mach-Zehnder Interferometer, (MZI), is the most adequate configuration for the monolithic fabrication of integrated optics microsensors. In fact, this structure is particularly suitable for sensing problems as it can combine high resolution and high sensitivity performances together with an excellent insensitiveness on mechanical vibration or other environmental effects. Many applications of optical sensors based on MZ interferometer have been reported in literature, both in the case of already cited Fourier Transform, (FT), and spectroscopy. In the case of Bio-chemical analyses, the phase shift between the electromagnetic waves propagating in the two arms of the MZI is generated by the adhesion of the analyte molecules on the surface of one (sensing) arm that has been left uncovered by the cladding protecting the whole device from the environment [4–6]. The working principle, in this case, is based on the optical path variation, $\Delta p = \Delta nL$, generated by the change of the refractive index in the uncovered sensing arm, having a length L. A relatively limiting factor of this instrumental architecture is originated from the need for a relatively large value of the sensing interferometric arm L, to have a good sensitivity of the device. Moreover, to obtain the best performances of an MZI in terms of sensitivity, requires working with monochromatic light, for this reason, it has been suggested [6] to design a hybrid detection architecture, specifically oriented to biochemical applications, allowing both high sensitivity and a high selectivity in the case of multi-analyte detection. This performance can be obtained by integrating an MZ interferometer with a spectrum analyser and using a multi-wavelength 'white' light source to distinguish the spectral 'signature' of the different species. **Figure 4** is redrawn the device originally suggested in Ref. [7].

As previously anticipated, when the monolithic interferometer is fabricated on Electro-optic material, typically $LiNbO_3$, the phase shift can be generated by exploiting the Pockels effect to modify the refractive index in one of the two arms, through the application of an electric field using two electrodes placed in a suitable position close to the optical waveguide (see **Figure 3(b)**).

If a continuous variation of the phase shift is generated, it is realised a so-called scanning interferometer and the whole set of intensity data measured as a function of the phase shifts gives rise to the so-called Interferogram. The Fourier Transform of the Interferogram gives rise to the frequency spectrum of the light containing the spectral information on the element to be detected [3, 4]. In the traditional instrument of **Figure 3(a)** the scanning effect is usually produced by uniformly moving one of the two mirrors giving rise to a corresponding optical path variation in one arm of the interferometer generating the desired phase shift variation. On the contrary, in an integrated microdevice, the scanning effect can be simply obtained by applying a voltage ramp to the electrodes (see **Figure 3(b)**), without the need for moving parts.

The detailed description of the Fourier Transform Spectroscopy principle and the detailed mathematical considerations have been extensively reported in a number of articles and textbooks [2, 3, 6]. In this work, we only report a concise description of the operation system of a scanning integrated MZI (see **Figure 3(b)**).

In order to achieve good spectral resolution, the scanning Interferometer needs phase shifts suitable to produce interferograms of many tens of interference fringes. This involves the need that the substrate material has an electro-optical coefficient as high as possible. In our case, the output intensity is monitored as a function of the optical path variation induced by a suitable variable electric field, applied to the arms. The interferometric output I_{out} represents the Fourier Transform of the input spectral distribution $I(k)$ and is given by the following relationship:

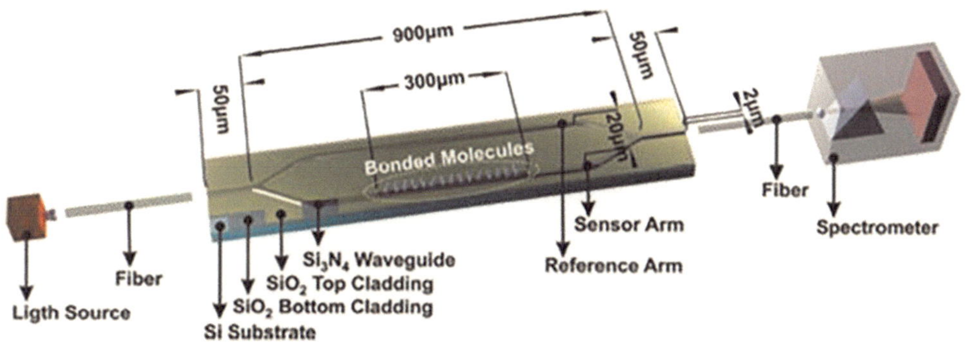

Figure 4.
Hybrid MZI with a sensing pad in one arm integrated with a wavelength dispersive system [7].

$$I_{out}(\Delta p) = \int I(k)\ [1 + \cos\,(2\pi\Delta p(k))]dk, \tag{1}$$

where k = 1/λ is the wavenumber of the incident radiation, and Δp is the optical path difference.

In our case, the optical path difference can be expressed by the linear relationship

$$\Delta p = L\ \Delta n, \tag{2}$$

where: L is the arm length, and

$$\Delta n = r_{33} \cdot n_e^{\,3} \cdot E/2, \tag{3}$$

where r_{33} is the linear electro-optic coefficient along the optical c-axis, n_e is the $LiNbO_3$ extraordinary refractive index when x or y propagation is considered. Finally, E is the electric field applied to the driving electrodes.

From Eq. (1) it appears that in the case of monochromatic light, the intensities recorded by the detector, (i.e. the Interferogram), has a sinusoidal shape. In the real case, we deal with quasi-monochromatic light and the MZI output are somehow distorted as reported in **Figure 5** in the case of the incident light generated by a diode laser emitting in a narrow window around 635 nm.

When dealing with a wideband ('white') light, the conceptual behaviour is the same. Even if the interferogram is remarkably modified and several corrections must be applied to the over-simplified scheme previously reported [3], it is always possible to perform the spectrometric analysis of the incoming light. **Figure 6** reports an example of a real case of E131 dye (Patent Blue) detection with an integrated MZI.

More recently, a new generation of integrated architectures has been suggested for spectroscopic analyses, in particular the integrated monolithic version of the Young interferometer (YI) and the integrated monolithic version of the Michelson's ('Echelle') Diffractometer. **Figure 7** reports the sketch of the Young geometry in the integrated version. In this case, the waveguides' geometries are geometrically arranged like in the MZI, the only difference is at the detection side that is here based on an arrayed detector.

A more recent architecture has been realised on the basis of Michelson's echelle grating that, when fabricated with electro-optic material ($LiNbO_3$), becomes a Programmable Micro diffractive Grating (PMDG) extending the range of this kind of micro-optical device to an extremely large field of applications. **Figure 8** reports the

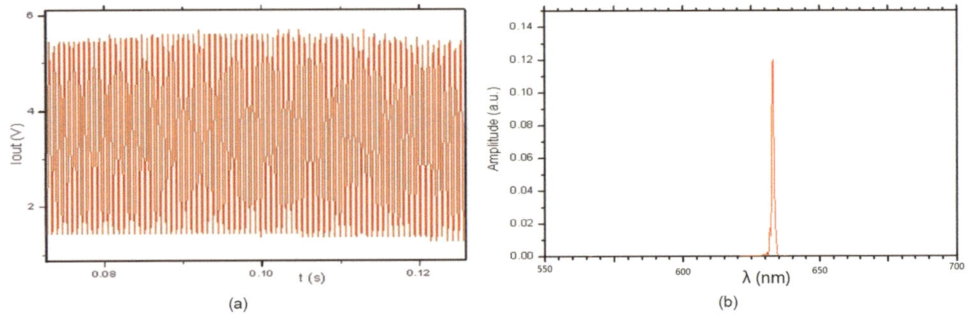

Figure 5.
(a) Raw signal detected during the scanning cycle, (interferogram) and (b) the Fourier transform of the interferogram, giving the wavelength spectrum of the incident light.

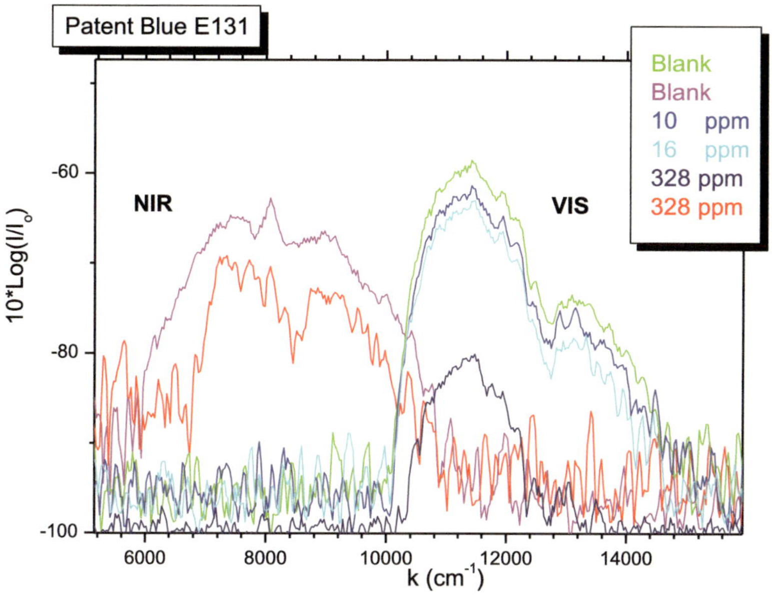

Figure 6.
Example of densitometry performed with an IO micro-interferometer on an E131 dye solution.

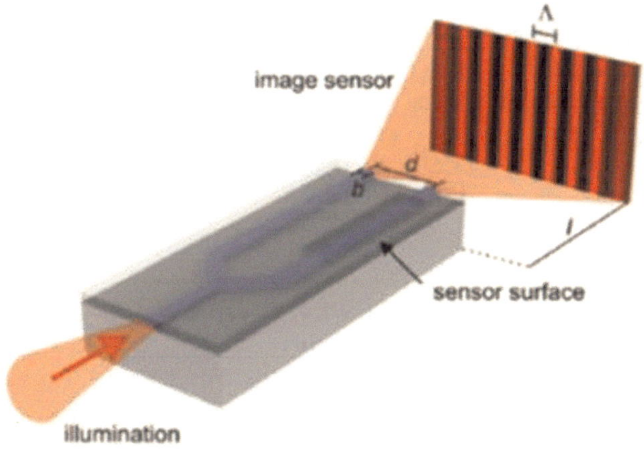

Figure 7.
Integrated version of the young interferometric geometry the far-field geometry for the detection of the modifications of the interference pattern, originated by the analyte adhesion on the sensing pad in one of the two arms.

comparison between the original static diffractive grating proposed by Michelson and the integrated PMDG, fabricated for high sensitivity detection systems based on the correlation spectroscopy technique. In Michelson's original device, the wavefront portions emerging from the different steps crosses different glass thicknesses, so the different optical paths acquire different phase shift as a function of the glass thickness crossed. The different wavefront emerging from each step interfere with each other giving rise to a far-field diffraction pattern that can be observed in far-field conditions on screen S, of **Figure 8(a)**.

If the integrated waveguides array sketched in **Figure 8(b)** is fabricated on electro-optical substrate, it can have the same behaviour as the Michelson's echelle device.

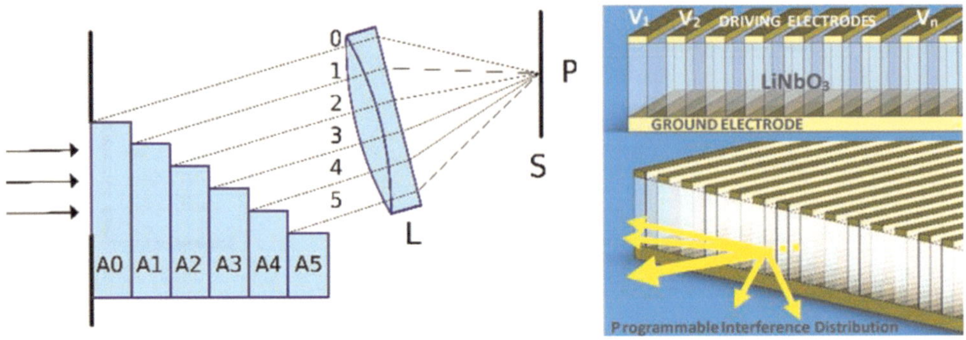

(a) (a)

Figure 8.
(a) Michelson's echelle diffraction grating. (b) Integrated programmable diffractive grating (PMDG) based on $LiNbO_3$ substrate.

In fact, each waveguide may have a different refractive index as a function of the electric field applied to the electrodes so, each portion of the wavefront emerging from each waveguide has a different phase and, in far-field conditions, they interfere with each other generating a diffraction pattern like in the case of Michelson's device.

The great difference between the two cases is that the device shown in **Figure 8(a)** is static and the diffractive properties are fixed by the construction parameters, whereas the PMDG device, the diffractive properties are programmable simply by changing the voltage applied at each electrode. This feature enormously expands the field of the applications of this Integrated Optics microdevice that can go far beyond the sensing systems arriving to have implications in many strategic areas such as optical fibre transmission, cryptography, quantum optic devices, optical computing, etc.

3.1 Holographic correlation spectroscopy with PMDG device

Correlation spectroscopy architectures have been widely studied for at least 20–30 years and are now argument described in the textbooks [8–10]. Therefore, in the present work, we will not enter in detail in the presentation of this technique and we will take as a reference the 'holographic correlation spectroscopy' architecture [11] treating in detail the subject of correlation spectroscopy in connection with the use of computer-generated optical elements. In particular, we will consider the conclusion of reference [11] when discussing the use of the PMDG to create synthetic spectra of several compounds, some of which are of interest for Environmental control, Food production and transportation, Bio-Chemical hazard, safety and security problems.

In particular, due to its programmability, the device can generate, at least in principle, synthetic spectra of almost any analyte of interest. The PMDG may have hundreds of diffractive elements in very small overall dimensions and, with the suitable software, it can generate a digital library containing the synthetic spectra of plenty of molecules. When the PMDG is exploited in this architecture, the sensing instrument described here can be considered as an example of a new sensor concept, in fact, in traditional instruments, the optical spectrum processing is determined by the correlation between the light transmitted through an unknown sample and the light transmitted through a reference cell containing a known mixture of the chemical molecule to be detected. In this case, due to the PMDG properties, the optical spectrum processing is based only on the correlation between the light transmission through an unknown sample and the data of a digital library. This trait greatly increases the sensor flexibility if compared with other recent instruments in which the spectrally dispersed light is delivered onto a coded mask to provide spectral filtering of the sample spectrum light [11].

In conclusion, this new sensor is particularly suited for safety and security applications because it avoids the use of a reference cell containing reference materials that can be difficult and hazardous to handle, in the case of detection of dangerous, poisoning, or explosive targets. Furthermore, it allows great flexibility if compared with the coded mask filtering described in Ref. 11, because the number of molecules that can be detected is now only limited by the wavelength transmission window of the electro-optic material used as a substrate for the PMDG device and by the sensitivity range of the detector used. Finally, exploiting during the data collection the dark-field correlation technique, it can be obtained a remarkable increase in the sensitivity of the whole system. In fact, by using shrewdness to construct the reference synthetic spectrum to be the complement of the target transmission spectrum, the signal-to-noise ratio becomes very large. Actually, with this shrewdness, all the wavelengths different from those matching the absorbance of the target, are blocked creating a programmable filter that allows the transmission of only a very small fraction of the incident light in correspondence of the desired wavelengths exclusively. So, when the spectral absorption lines match the planned transmission complement it is obtained a very large signal-to-noise ratio, allowing extremely high sensitivity detection.

The architecture of 'holographic correlation spectroscopy' is schematically reproduced in **Figure 9**, is particularly simple and takes advantage of the intrinsic PMDG flexibility and reconfigurability features. In the geometry of **Figure 9**, a broadband 'white' light coming from an external source crosses, (one or several times), a sample cell containing the analyte to be investigated. Then the radiation crosses the PMDG optical element, placed in transmission architecture. In the present case, the external broadband light source could cover the whole transparency range of the $LiNbO_3$ base material ranging from 0.450 μm to 5.5 μm.

The wideband radiation coming from the external source is then transmitted through the PMDG device and precisely defined by the ensemble of the programmed driving electrical potentials applied to the different waveguides. The degree of correlation between the sample cell spectrum, (unknown), and the synthetic spectrum

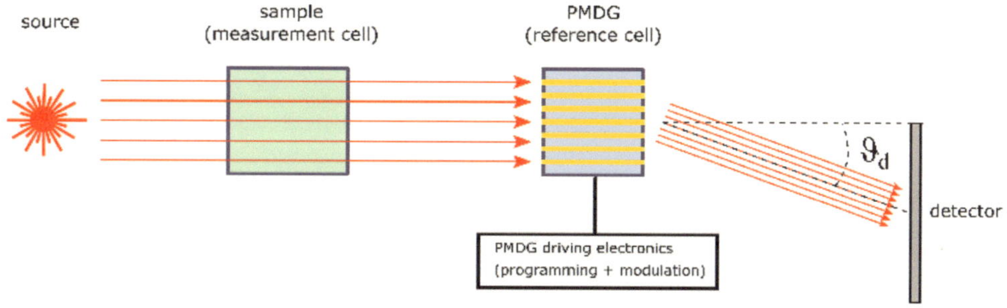

Figure 9.
Schematic of a holographic correlation spectrometer set-up including the PMDG which acts as the reference cell. The light coming from a broadband source passes through the measurement cell which contains the unknown analyte. The diffraction pattern generated by the PMDG at the diffraction angle ϑd is perfectly matched to the spectrum of the target compound within the measurement cell.

programmed through the reference (PMDG) cell is then measured on a photodetector placed at a suitable diffraction angle ϑ_d.

To increase the sensitivity and the selectivity of the set-up, the usual techniques can be adopted, in particular: (i) Lock-in techniques are applied by modulating the transmission spectrum through the PMDG and recording the detector output at the same modulation frequency or/and (ii) Darkfield technique, i.e. applying a suitable map of the driving electrical potentials, the PMDG can be configured to synthetize the complement of the real target to be analysed, so obtaining a much larger signal-to-noise ratio and a consequent remarkable increase of both sensitivity and selectivity of the detection system.

3.2 Synthetic spectra generation mathematical approach

The key point in the programming of a waveguide-based PMDG for a synthetic spectra generation is the determination of a driving voltage pattern able to introduce the required phase shifts on each of the M waveguides, (Typically M can range from 50 to 200 waveguide/cm). In this way, the electromagnetic radiation emerging from the output face of the grating will generate the desired synthetic spectrum at a predetermined diffraction angle ϑ_d.

In this section, a mathematical framework for the specific case in which the functional elements are electro-optical waveguides is presented and discussed [12].

Under the hypothesis of working in the Fraunhofer approximation, the diffracted field at an angle ϑ_d, (see **Figure 9**) can be described by the following Fourier-transform integral:

$$U(\vartheta_d, \lambda) = \int_{-\infty}^{+\infty} \int_{\lambda_1}^{\lambda_2} \frac{CA}{\lambda} U'(x, \lambda)\, exp\left(-\frac{i2\pi\, sin\,(\vartheta_d)}{\lambda}\right) d\lambda dx \tag{4}$$

where C is a constant of proportionality, A is the amplitude of the incident wave, supposed to be independent of the wavelength λ and x is the spatial transverse coordinate of the PMDG output facet whereas, λ_1 and λ_2 are the lower and upper limit of the spectral band of interest, respectively (**Figure 10**).

In Eq. (4), $U'(x, \lambda)$ is the near-field distribution emerging from the output facet of the PMDG. For a grating with M waveguides of width W and periodicity P, this distribution can be described by the following expression:

$$U'(x, \lambda) = \sum_{m=1}^{M} exp\left(\left(-\frac{i2\pi L}{\lambda}\right)\left(\frac{Vm\gamma}{d}\right)\right)\, exp\,\left(-\frac{(x - mP - P/2)^2}{\left(\frac{W}{2}\right)^2}\right) \tag{5}$$

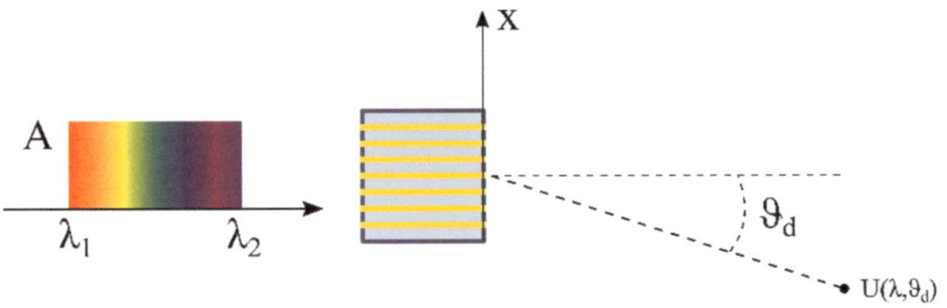

Figure 10.
Framework for the solution of the PMDG diffraction problem.

where L is the electrode length, d is the device thickness, V_m is the voltage applied to the *m-th* waveguide and γ is a constant depending both on the electro-optical characteristics and the crystallographic orientation of the substrate. In Eq. (5), the first exponential term of the summation describes the phase shift induced on the *m-th* waveguide by the local driving voltage V_m, whereas the second exponential term describes the mode profile emerging from each waveguide that, as a first approximation, is supposed to follow a Gaussian shape. Eq. (5) clearly shows how the local phase of the field emitted by the grating and, consequently, the diffracted pattern at angle ϑ_d, can be controlled through the application of a suitable set of M potentials V_m=1, ... , M. The PMDG design and programming procedure is therefore reduced to the determination of a suitable voltage pattern, which minimises the difference between the target spectrum of interest and the synthetic one calculated with Eqs. (4) and (5). The difference between the target and the synthetic spectra can be quantified by introducing an error function $\mathcal{E}$ of the form:

$$\mathcal{E} = \sqrt{\sum\nolimits_{n=1}^{N} \left(I_n^{\,d} - I_n\right)^2} \tag{6}$$

where $I_n = |U(\lambda_n)|^2$ is the intensity of the target spectrum and I_n^d is the diffracted intensity spectrum at the considered diffraction angle ϑ_d, both evaluated over the same set of N wavelength within the spectral range of interest. Once this error function has been defined, the PMDG programming procedure reduces to the optimization problem of finding the minimum of $\mathcal{E}$ with respect to the control parameters (V_1, V_2, ... V_M). By setting $K = L\gamma/d$ we can write, $D_m = K\,V_m$ and the new variables (D_1, D_2, ... D_M) have the dimension of a length, so each D_m value considers the optical path differences, the different waveguides.

This is physically equivalent to introducing a programmable phase-shift offset in the first term of Eq. (5). Then, the actual voltage pattern to apply to the different electrodes of the PMDG is calculated from the D_m values, resulting from the optimization routine, once the technological parameters have been defined.

For the solution of the multivariable optimization problem, several numerical approaches have been proposed and implemented so far, in particular, iterative Fourier Transform phase-retrieval algorithm [12], genetic algorithms [11] or gradient-based multi-variable minimization routines [13]. The Nelder–Mead Simplex Method [14] demonstrates to be extremely effective providing a monotonic and rapid convergence to the minimum of the error function $\mathcal{E}$. Moreover, the numerical implementation of this method is available in the most common scientific computation libraries.

An example of synthetic spectrum created with the numerical method of reference [14] is reported in **Figure 11** where a portion of the $COCl_2$ spectrum has been reconstructed in the hypothesis of a PMDG having M = 50 waveguides W = 8 μm and pitch P = 80 μm. The experimental absorbance spectrum is shown in **Figure 11(a)**, whereas the calculated synthetic one at the optimal diffraction angle ϑ_d of 3.75 degrees is presented in the (b) panel of the same Figure. The (c) panel shows the optimised parameters set D_m calculated with the numerical routine that, once introduced in (2), generates the synthetic spectrum displayed in (b). The agreement between the two spectra is remarkable over the entire wavelength range of interest, confirming both the effectiveness of the PMDG device and of the numerical optimization routine adopted to retrieve the programming pattern.

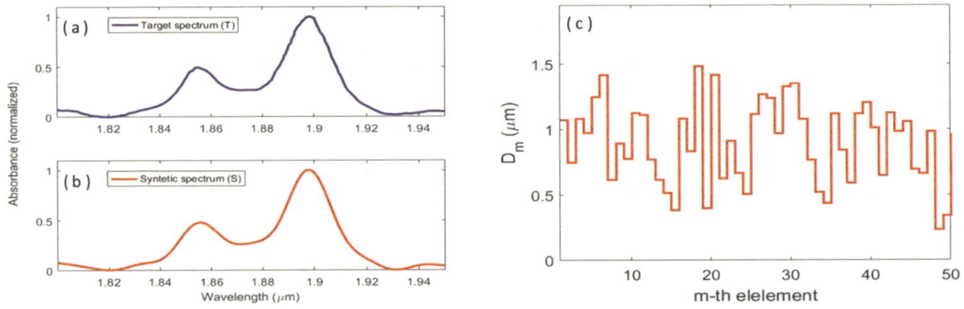

Figure 11.
(a) Experimental absorbance spectrum of gaseous phosgene (target spectrum T) and (b) PMDG-synthetized spectrum of the same analyte (synthetic spectrum S). (c) Values of the optimised D_m (m = 1. .,50) control parameters.

4. Integrated interferometers for ghost imaging in the spectral domain

The Ghost Imaging, (GI) phenomenon is based on the spatial correlation of light to form images and, since the early pioneering work in 1995, several experiments on the argument have been presented [15–18]. The GI is obtained by correlating the intensities of two spatially separate light beams, one of the two light beams illuminates an object to be imaged. The second beam, which does not see the object, (reference beam), if it is detected with a position-sensitive sensor, gives rise to the spatial image of the object created by the non-interfering photons.

Besides the GI in the space domain, another kind of 'ghost experiments' were performed in the frequency (Spectral), domain [19–21]. It must be pointed out that in Ref. to [20], thes kinds of experiments were performed using a classical thermal light emitted by a broad-band superluminescent diode. These experiments allowed to exploit real spectroscopy measurements detecting spectral lines of chemical molecules like $CHCl_3$ [20] and the spectral lines of Er^{3+} dopant [21] in $LiNbO_3$ non-linear material placed in one arm of an Asymmetric Non-Linear Interferometer, (ANLI). These preliminary approaches to experimental spectroscopic sensing demonstrate that a new field of sensing can be opened exploiting effects such as 'ghost spectrometry'. These new sensing techniques can allow extending the sensing limits of the traditional detectors and performing spectrometric measurements with non-interacting photons.

Before describing the experimental procedures, of ghost imaging in the frequency domain, let us give a rapid oversimplified description of the basic principles of the 'ghost' phenomena. In Ref. [22], this architecture was studied in detail and it was concluded that it gives rise to a condition that is often referred to as maximally entangled states of a two-mode field. If the two modes are physically separated like in the case of the arms of an MZI, we are in presence of separate path-entangled states that can have important applications to interferometry and interferometric analyses.

In particular, considering the case of a coherent intense light beam injected in an integrated MZI, the photons are spatially confined in geometrically separated single-mode waveguides, where they have a very high spatial and time coherence [22]. Moreover, the photon density inside the waveguides can reach values that are orders of magnitude higher than in the case of free space propagation. In these conditions, an integrated optical device is particularly suitable for both practical applications and basic quantum optics studies.

The integrated version of an ANLI, used to obtain ghost imaging in the frequency domain [23], is schematically reported in **Figure 12** as taken from reference [21].

In that case, the ANLI was not built as a monolithic device but was realised in a hybrid set-up in which the 980 nm Laser source was injected in a 50:50 Y coupler single-mode optical fibres. One arm was coupled with an Er^{3+} doped $LiNbO_3$ monolithic waveguide, whereas the other arm was coupled with a monolithic Pockels phase modulator as the one shown in **Figure 2**. Then the MZI geometry was completed by coupling the two arms with an identical second optical fibre Y coupler. Finally, the injected 980 nm photons arrive in the photodetector passing through a (975 ± 25) nm pass-band filter that eliminates all the photons generated by the interaction of the 980 nm pump with the Er^{3+} doped nonlinear $LiNbO_3$ arm. In any case, the photons generated by the interaction in the doped arm can only give rise to a continuous background because they cannot contribute to the interference process due to the 'Which Way' criterion.

The principle of the experiment was based on the Quantum properties stating that two Fourier Transform pairs are conjugate variables. So, operating a Fourier Transform in the time-frequency space-domain, it is possible to get information on the wavenumber space-domain of the 930 nm pump photons annihilated, (generating up or down-conversion), as a consequence of the pump photons interaction with the Er^{3+} doped nonlinear $LiNbO_3$. This intriguing effect takes place when the doped material is placed in one arm of the ANLI and the photodetector do not even see the photons generated by up or down-conversion in the doped arm.

The spectroscopy experiment was performed by injecting the 980 nm laser beam in the input port '*a*' of **Figure 12**. A fraction of the pump 980 nm photons is absorbed by the Er^{3+} dopant, giving rise to a series of up and down conversions having a widely studied spectral composition as reported in literature. Then, a linear voltage ramp is applied to the undoped arm of the ANLI producing a continuous phase shift variation between the photonic states entangled over the two optical paths of the ANLI. The pass-band filter ensured that only the 980 nm pump photons could reach the detector. The experimental data were collected detecting the intensity of transmitted photons as a function of time (i.e. of the phase shift), giving rise to the raw interferograms

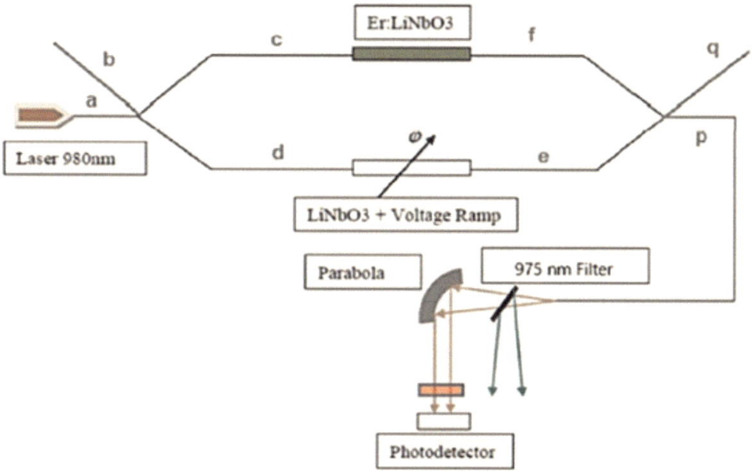

Figure 12.
Schematic of the experimental layout. The ANLI is used in [21].

shown in the inserts of **Figure 13(a)** and **(b)**. Then, the raw interferograms were elaborated by the usual methods used in Fourier Transform Spectroscopy [18, 24]. The same measurement was first performed by using an InGaAs photodetector having a detectability window extending up to wavelengths of 2.5 μm (in the SWIR region). The measurement was then repeated using a Si p-i-n detector that is 'blind' at wavelengths $\lambda \geq 1.1$ μm. The results of the two experiments are reported in **Figure 13(a)** and **(b)**, respectively.

In **Figure 13**, the two spectroscopic measurements of the Er^{3+} energy levels performed with an In GaAs *p-i-n* photodetector, (**Figure 13(a)**) and with a Si *p-i-n* photodetector, (**Figure 13(b)**) are comparatively reported.

The complete Er^{3+} Spectrum extending form visible to SWIR wavelengths was obtained reporting most of the spectral lines as listed in the literature.

The difference in the amplitudes can be attributed to the different sensitivities of the two detectors. Moreover, as previously clarified, only the 980 nm monochromatic pump photons that have not interacted with the Er^{3+} doped crystal can reach the detector. In fact, the 980 nm photons that interacted with Er^{3+} are annihilated, giving rise to the up or down-conversion photons that are eliminated by the (975 ± 25) nm passband filter.

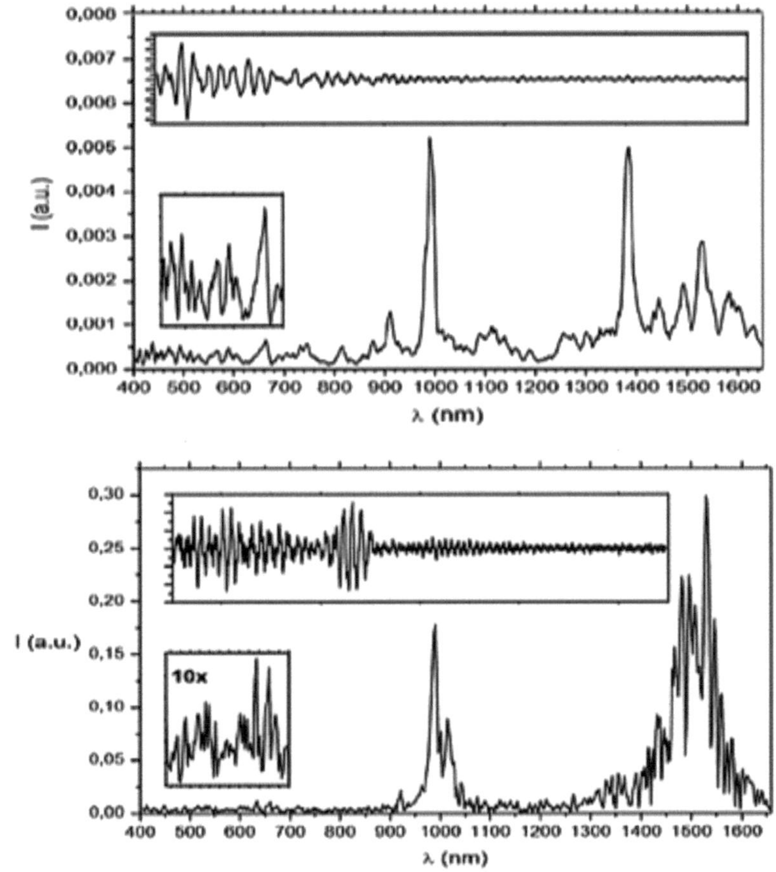

Figure 13.
Spectra obtained by Fourier transform analyses with the setup of ***Figure 12*** *(see text).*

In conclusion, with the ANLI integrated architecture reported in **Figure 12**, the Er^{3+} spectral lines appeared independently of the different sensitivity of the used detectors also over the whole SWIR, where the Si *p-i-n* detector is 'blind'. Considering that, due to the passband filter, only the non-interacting monochromatic photons have reached the detectors, we can say that the Er^{3+} spectral lines are 'ghost' lines produced by the separate path entangled states correlation [22] generated in the two arms of the interferometer.

The previously described results give a strong indication that, by using 'ghost imaging' in the frequency domain, it is possible to develop a new generation of integrated interferometric instruments, (most likely based on the ANLI architecture of **Figure 12**), having the capability to work with interaction-free photons. This allows extending the spectral measurement remarkably beyond the photodetector sensitivity limits, in particular in spectral regions where the photodetectors are not available or have too low sensitivity. In particular, some preliminary contacts indicate that this effect is of strong interest for several applications, with particular attention to astrophysical applications.

5. Integrated optics interferometry for astrophysics

5.1 Stellar interferometry with integrated optics

Optical long baseline interferometry is a technique that is undoubtedly providing high angular resolution observations in optical astrophysics. Fizeau in 1868 [25] was the first to attempt using interferometry for astronomical observations, without reaching the wanted result, eventually proposed and revised by Stéphane in 1874 [26]. Nevertheless, only in 1921, Michelson and Pease [27] first succeeded in measuring stellar diameters with a single telescope equipped with a pupil mask. The schematics and characteristics of their apparatus are reported in **Figure 14**. Unfortunately, their interferometer was not that sensitive to allow further investigation. In fact, a 1.0 milliarcsecond diameter on the sky translates to 0.5 μm in optical path difference (OPD), on a $B = 100$ m baseline (see **Figure 14**).

In practice, modern direct interferometry only started in 1975 with Labeyrie [28] who produced stellar interference with two separated telescopes.

Modern long baseline interferometry requires the combination of several stellar beams collected from different apertures (telescopes). The first interferometers started working with only two apertures, such as GI2T [29], SUSI [30], PTI [31], IOTA [32], COAST [33] and NPOI [34, 35]. The increase in the number of apertures was one of the major features of new generation interferometers. Today, the principal operational interferometric observatories, which use this type of instrumentation, include VLTI [35–45], and CHARA [37].

Current projects are using interferometers to search for extrasolar planets, either with nulling techniques, by astrometric measurements of the reciprocal motion of the star or through direct imaging.

In **Figure 15**, the geometry of the ideal interferometer is reported. Let us specify the incident source flux power F in units of energy incident per unit time per unit cross-sectional area, and the collecting area of the apertures A_1 and A_2 as A. Then, apart from some efficiency factors, the detected power P expresses as per the following:

$$P = 2AF\,(1 + \cos k(s\ \ B + d_1\ \ d_2)), \tag{7}$$

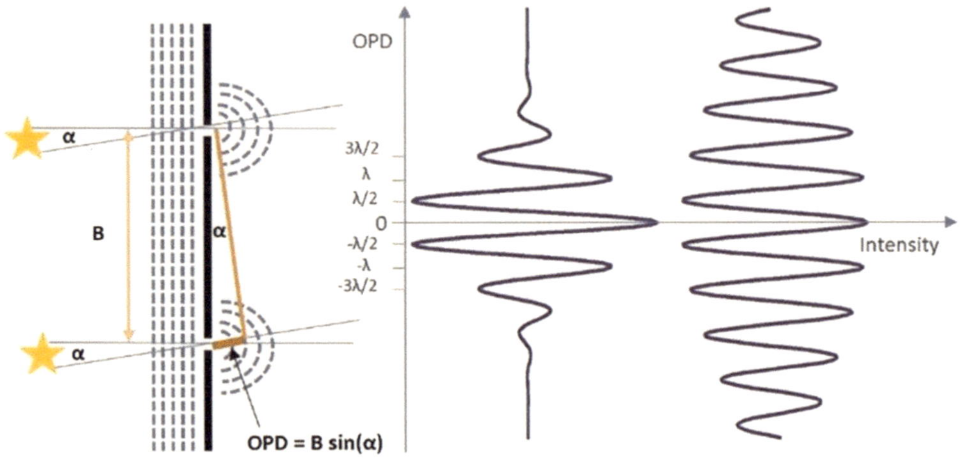

Figure 14.
(a) Scheme of the two slits mask experiment from Michelson and Pease. A star, with an α angular diameter, is imaged after its light passes through a double slit mask, with B as the slit distance. An interferogram appears as a function of the optical path difference (OPD), with the first minimum at OPD = λ/2, where λ is the wavelength of the impinging light. The fringes disappear when the OPD overcomes the source coherence length, e.g. (b) and (c) show different coherence length interferograms.

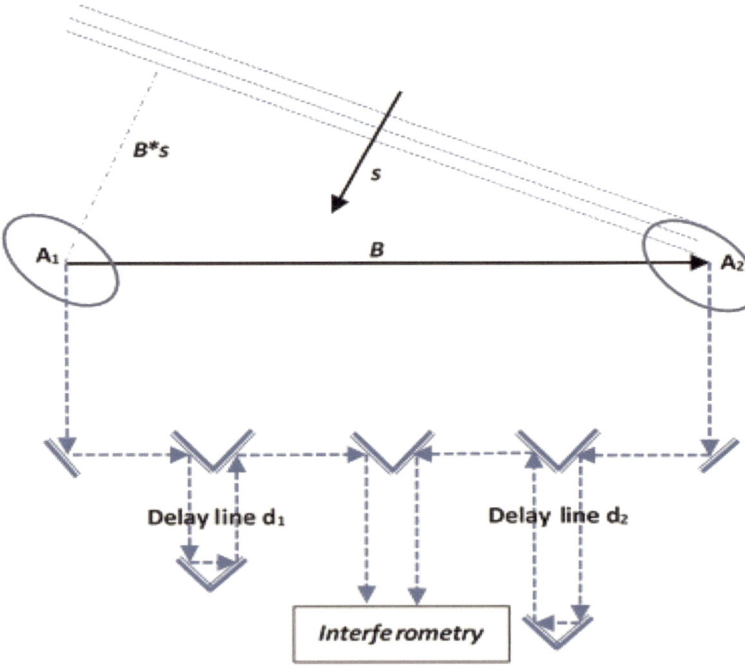

Figure 15.
Ideal stellar interferometer schematics.

where $k = 2\pi/\lambda$, $\boldsymbol{s}$ is the unit vector normal to wavefront propagation and $\boldsymbol{B}$ is the baseline vector between the two apertures.

In the space of relative delay $D = \boldsymbol{s}^* \boldsymbol{B} + \boldsymbol{d_1} * \boldsymbol{d_2}$, P varies harmonically between zero and $2AF$ with period λ. It is important to recall that for conventional imaging, the limiting angular resolution α follows the well-known relationship $\alpha \sim \lambda/D$, (where D is the telescope pupil diameter and λ is the wavelength), therefore, in comparison, an interferometric system provides the measure of interference fringes between two beams at higher angular frequencies, in the order of $\alpha \sim \lambda/B$.

The complex visibility of these fringes is proportional to the Fourier transform of the object intensity distribution (Van-Cittert Zernike theorem), hence allowing to resolve particulars, very narrow from the angular point of view. Following these principles, stellar interferometry is offering to present days astronomers the ability to study celestial objects in unprecedented detail. It is possible to see details on the surfaces of stars and even to study celestial bodies close to a black hole [46] (**Figure 16**).

Stellar interferometry has become even more effective due to the advent of high sensitivity detectors and of large aperture telescopes. Nevertheless, to implement it, a complex system of mirrors has normally to be set up to bring the light from the different telescopes, constituting the synthetic aperture, to the instruments, where it is combined and processed (see scheme in **Figure 17**). This is technically demanding,

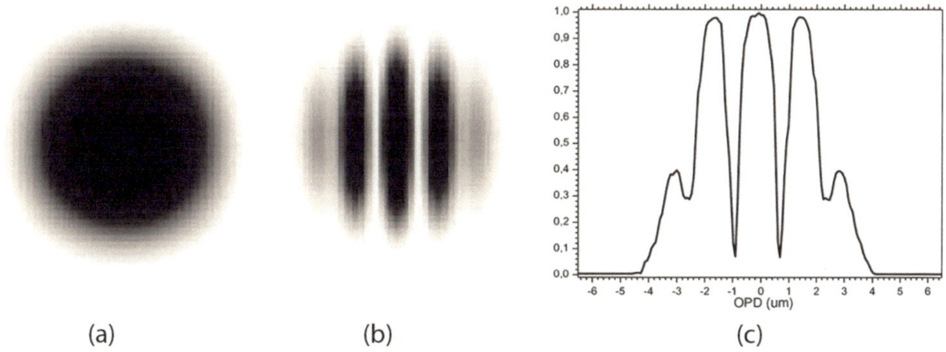

Figure 16.
(a) Conventional star image, (b) same with two telescopes stellar interferometry having baseline B = 10D, *and, (c) corresponding interference pattern profile.*

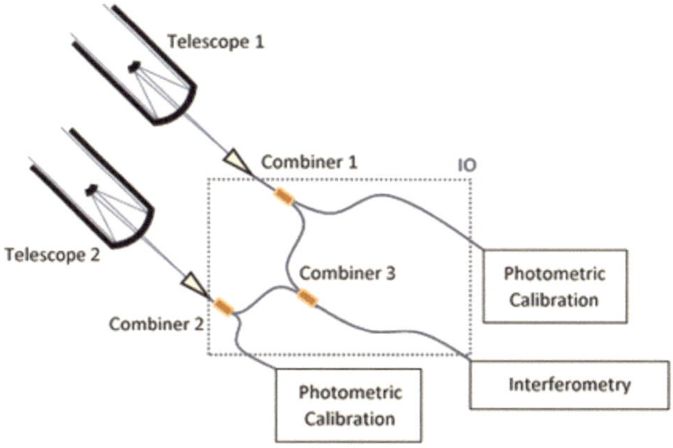

Figure 17.
Schematics of IO apparatus required for two telescopes stellar interferometry. Light from the same source collected by telescopes 1 and 2 is injected in 2x2 IO beam splitters/combiners 1 and 2, respectively. One output port of each combiner is used to perform photometric adjustment whereas the second ports are combined in a third IO 2x2 beam splitter/combiner, at which output ports the interferometric signal is collected and processed.

(a) (b)

Figure 18.
(a) Fibre pig-tailed IO y-branch mounted on mechanical support, (b) SEM picture of the IO Y-junction, allowing beam splitting as well as fibre signal combining.

as the light paths have to be set equal to within 1.0 μm over distances of a few hundred metres, in order to avoid an OPD offset out of the coherence length.

For ground-based interferometers, the source phase is corrupted by atmospheric turbulence. This prevents an absolute measurement of the source phase. However, it is possible to measure the difference in the source phase between two wavelengths. In practice, stellar interferometry requires star tracking techniques to compensate for astronomic seeing due to atmospheric turbulence.

In recent years, integrated optics and photonics technology, inherited from the telecom field and micro-sensor applications, was proposed for astrophysical interferometry. Results obtained with components coming from micro-sensor application were first presented by Berger et al. in a seminal series of dedicated works [38–44]. These authors demonstrated the validity and feasibility of the integrated optics technology for astronomical interferometry, by using telecom fibre coupler/combiners. Following a complete laboratory characterisation of the optical properties of the applied IO components, a first set-up was tested at the Infrared Optical Telescope Array (IOTA) observatory, in Arizona.

The above-mentioned studies demonstrated that beam combiners are very stable and lead to precise measurements. Moreover, IO components are versatile and easy to handle. In particular, the number of optical alignment adjustments strongly simplifies, which dramatically reduces the complexity of multiple-beam combinations for aperture synthesis imaging (**Figure 18**).

Other examples of IO based stellar interferometers are present by the VLTI, where the VINCI apparatus, based on IO beam combiners and fibre optics components, has allowed astronomers to reach the unprecedented resolution of 4.0 milliarcseconds in sky observations [46].

5.2 IO Mach-Zehnder micro-interferometers for earth and space remote sensing

Absorption or emission spectroscopy is largely adopted for remote sensing in both Earth and Space exploration, on board dedicated satellite platforms.

In this case, all general resources (weight, encumbrance, energy consumption, etc.) are particularly limited, furthermore, the onboard instrumentation is exposed to harsh environmental conditions (vibrations, ultra-high vacuum, radiation, temperature gradients, etc.). For these reasons, IO devices can represent a very important solution, particularly when based on monolithic structures (**Figure 19**).

Integrated scanning micro-interferometers with Mach–Zehnder geometry, have been designed and produced by using MEOS (Micro Electro Optical Systems) technologies.

DOI: http://dx.doi.org/10.5772/intechopen.103770

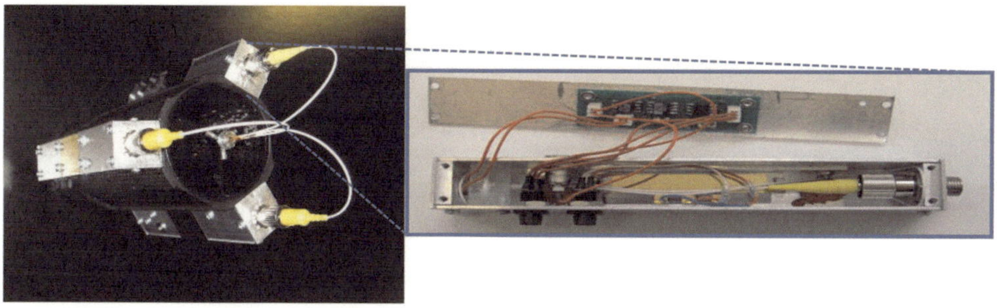

(a) (b)

Figure 19.
(a) Carbon fibre telescope integrating three IO micro-interferometers. (b) Detail of an integrated MZ micro-interferometer equipped with front-end optics, readout electronics and packaging, ready to use (the overall package length is 12 cm).

The obtained micro-devices are based on integrated optical waveguides on $LiNbO_3$ (LN) crystals, electrically driven, without moving parts, by exploiting the electro-optical properties of the material. These IO devices are Fourier Spectrometers in that they operate the Fourier Transform of the input radiation spectral distribution, which is eventually recovered starting from the output signal by means of Fast Fourier Transform (FFT) techniques.

Such micro-interferometers weigh a few grams, require a power consumption of a few mW and, in principle, can operate in the whole LN transmittance range (0.36 μm–4.5 μm).

In the literature several works have been reported [12, 47, 48] describing the development of a whole series of micro-interferometric apparatuses, demonstrating in principle the applicability of IO MEOS technology for Space exploration and Earth remote sensing. The micro-interferometers were produced on x-cut $LiNbO_3$ single-crystal substrates, by applying non-conventional micromachining techniques, based on high-energy particle beams processing.

Performances were particularly tested in the 0.4 μm–2.5 μm spectral window, with some extension also in the 2.5 μm–4.5 μm range. In the Visible region 0.4 μm–0.7 μm this microsystem demonstrated a spectral resolution suitable for detecting the

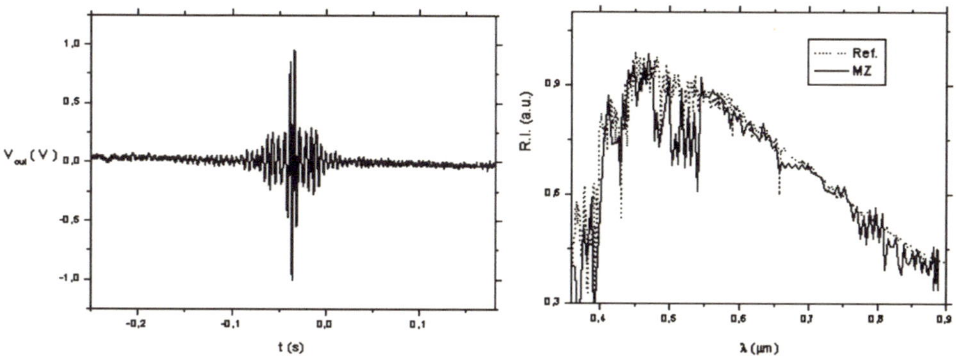

(a) (b)

Figure 20.
(a) Solar radiation interferogram, (b) corresponding FFT (solid), reference (dot).

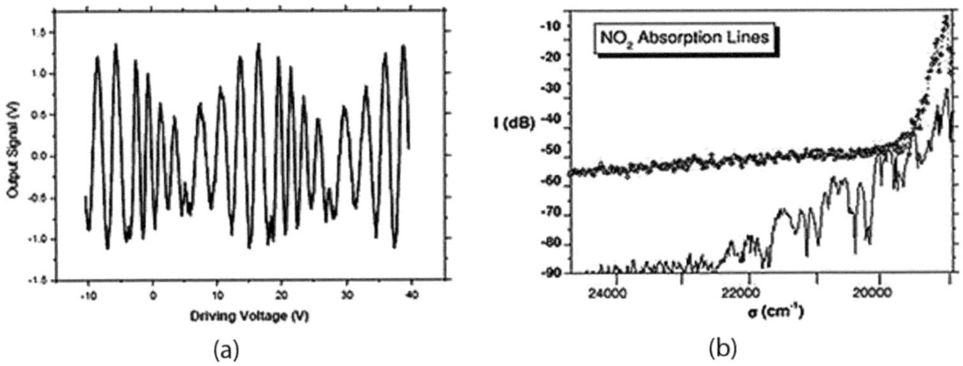

Figure 21.
(a) Raw interferogram as obtained from an integrated scanning MZI, and (b) the absorption analysis of the NO_2 analyte, (lower curve), introduced in a wide band light (upper curve).

characteristic lines of the solar spectrum together with the absorption bands of common gases present in Earth's atmosphere (see **Figures 20** and **21**).

6. Conclusions

Recent developments of Integrated Optics and Photonics components allow implementing complex, rugged, robust and miniaturised interferometric systems, which applications span all the fields of sensing interferometry also including the possibility of upcoming quantum effects applications, already shortly mentioned in paragraph 4. In this work, a general overview of the present state of the art was presented, with the intention to stimulate the interests of investigators and researchers operating in multidisciplinary contexts and in general in all fields of sensing and analytics.

Author details

Marco Chiarini[1*], Alberto Parini[1] and Gian Giuseppe Bentini[1,2]

1 Prometheus Srl, c/o CNR-IMM, The Italian National Research Council, Institute for Microelectronics and Microsystems, Bologna, Italy

2 CNR-IMM, The Italian National Research Council, Institute for Microelectronics and Microsystems, Bologna, Italy

*Address all correspondence to: chiarini@bo.imm.cnr.it

References

[1] Kozna P et al. Integrated planar optical waveguide interferometer biosensors: A comparative review. Biosensors and Bioelectronics. 2014;**58**:287-307. DOI: 10.1016/j.bios.2014.02.049

[2] Born M, Wolf E. Principles of Optics. Cambridge, UK: Cambridge University Press; 1999. DOI: 10.1017/CBO9781139644181

[3] Fritzsche W, Popp J. Optical Nano- and Microsystems for Bioanalitics in Springer Series on Chemical Sensors and Biosensors. Germany: Springer Verlag Berlin Heidelberg; 2012. pp. 103-154. DOI: 10.1007/978-3-642-25498-7_4

[4] Kuntz RE. Miniature integrated optical modules for chemical and biochemical sensing. Sens. & Actuat. 1997;**B38**:13-28. DOI: 10.1016/S0925-4005(97)80167-0

[5] Heideman RG et al. Performance of a higly sensitive optical waveguide Mach-Zehnder interferometer immunosensor. Sens. & Actuat. 1993;**B10**:209-217. DOI: 10.1016/0925-4005(93)87008-D

[6] Kitsara M et al. Broadband Mach-Zehnder interferometry as a detection principle for label-free biochemical sensing. In: SENSORS Proceedings; 26–29 October 2008; Lecce, Italy: IEEE; 2008. pp. 934-937. DOI: 10.1109/ICSENS.2008.4716594

[7] Wilson J, Hawkes J. Optoelectronics: An Introduction. Prentice Hall NJ; 1998

[8] Zhou G et al. Design of the diffractive optical elements for synthetic spectra. Optics Express. 2003;**11**:1392-1399. DOI: 10.1364/OE.11.001392

[9] Sinclair MB et al. Correlation Spectrometer. 2010. U.S. Patent 7 697 134 Bl

[10] Smith MW et al. Theoretical description and numerical simulations of a simplified Hadamard transform imaging spectrometer. In: Proceedings of International Symposium on Optical Science and Technology, 8 November 2002. Vol. 4816. Seattle, WA, United States: SPIE; 2002. pp. 372-380. DOI: 10.1117/12.451698

[11] Basti G et al. Sensor for security and safety applications based on a fully integrated monolithic electro-optic programmable micro diffracting device. In: Proceedings SPIE Security + Defence, 2019, Strasbourg, France. Electro-Optical and Infrared Systems: Technology and Applications XVI. Vol. 11159. Strasbourg, France; 2019. p. 1115907. DOI: 10.1117/12.2532107

[12] Bentini GG et al. Fully integrated Mach-Zehnder Microinterferometer on Lithium Niobate as an example of Micro electro Optical system for space applications. In 5th ESA round table on micro/nano technologies for space, Norvijk, The Nederlands, 2-5 October 2005. Available on line: http://www.congrex.nl/05c12

[13] Sinclair MB. Synthetic spectra: A tool for correlation spectroscopy. Applied Optics. 1997;**36**:3342-3348. DOI: 10.1364/ao.36.003342

[14] Bentini GG et al. Monolithic fully integrated programmable microdiffraction grating based on electro-optical materials. In: Proceedings of Proc., Photonic Materials, Devices, and Applications II, 12 June 2007. Maspalomas, Spain: SPIE; 2007. pp. 6593-659325. DOI: 10.1117/12.724171

[15] Pittman TB et al. Optical imaging by means of two-photons quantum entanglement. Physical Review A. 1995;

52(5):R3429-R3432. DOI: 10.1103/PhysRevA.52.R3429

[16] Markie D, van Voorthuysen EH. Realization of an interaction -free measurement of the presence of an object in a light beam. Am. J. of Phys. 1996;**64**: 1504-1507. DOI: 10.1119/1.18413

[17] Barreto Lemos G et al. Quantum imaging with undetected photons. Nat. Lett. 2014;**412**:409-412. DOI: 10.1038/nature13586

[18] Amiot C et al. Supercontinuum Spectral-domain ghost imaging. Optics Letters. 2018;**43**(20):5025-5028. DOI: 0.1364/OL.43.005025

[19] Scarcelli G et al. Remote spectral measurements using entangled photons. Applied Physics Letters. 2003;**83**:5560-5562. DOI: 10.1063/1.1637131

[20] Janassek P et al. Ghost spectroscopy with classical thermal light emitted by a superluminescent diode. Phys. Rev. Applied. 2018;**9**:021001-1-021001-4. DOI: 10.1103/PhysRevApplied.9.021001

[21] Chiarini M et al. Theoretic approach to ghost imaging in the frequency domain performed by means of a high brilliance coherent monochromatic source. In: Proceedings of SPIE Security + Defence, 2019, Strasbourg, France SPIE (2019) 11159, Electro-Optical and Infrared Systems: Technology and Applications XVI. 1115905. Strasbourg, France. DOI: 10.1117/12.2532307. Available from: https://www.spiedigitallibrary.org/conference-proceedings-of-spie/browse/SPIE-Security-Defence/2019

[22] Gerry CC et al. Nonlinear interferometer as a resource for maximally entangled photonic states: Application to interferometry. Physical Review A. 2002;**66**:013804. DOI: 10.1103/PhysRevA.66.013804

[23] Chiarini M et al. Spectroscopy of non interfering photons through a non-linear integrated optics Mach-Zehnder interferometer. In: Proceedings of Optics + Optoelectronics, 2017, Prague. Czech Republic. SPIE. 2017;**10230**:1023004-1023003. DOI: 10.1117/12.2265742

[24] Li J et al. Miniature Fourier transform spectrometer based on wavelength dependence of half-wave voltage of a LiNbO3 waveguide interferometer. Optics Letters. 2014;**39**: 3923-3926. DOI: 10.1364/OL.39.003923

[25] Fizeau H. C. R. Acad. Sci. Paris. 1868; **66**:932. DOI: N/A

[26] Stéphan E. C. R. Acad. Sci. Paris. 1878;**78**:1008. DOI: N/A

[27] Michelson AA, Pease FG. The Astrophysical Journal. 1921;**53**:249. DOI: N/A

[28] Labeyrie A. The Astrophysical Journal. 1975;**196**:L71-L75. DOI: N/A

[29] Mourard D et al. The GI2T interferometer on plateau de Calern. A& A. 1994;**283**(2):705-713. DOI: N/A

[30] Davis J et al. The Sydney university stellar interferometer -II. Commissioning observations and results. Month. N. of the Royal Astr. Soc. 1999;**303**(4):783-791. DOI: 10.1046/j.1365-8711.1999.02270.x

[31] Colavita MM et al. ASEPS-0 testbed interferometer. In: Breckinridge JB, editor. Amplitude and Intensity Spatial Interferometry II. Vol. 2220. SPIE; 1994. p. 89. DOI: 10.1117/12.177230

[32] Carleton NP et al. Current status of the IOTA interferometer. In: Breckinridge JB, editor. Amplitude and Intensity Spatial Interferometry II. Vol. 2200. SPIE; 1994. p. 152. DOI: 10.1117/12.177236

[33] Baldwin JE et al. The first images from an optical aperture synthesis array: Mapping of Capella with COAST at two epochs. A&A. 1996;**306**:L13

[34] Benson JA et al. Multichannel optical aperture synthesis imaging of zeta1 URSAE majoris with the navy prototype optical interferometer. Astr. J. 1997;**114**:1221-1226. DOI: 10.1086/118554

[35] White NM et al. Progress report on the construction of the navy prototype optical interferometer at the Lowell Observatory. In: Breckinridge JB, editor. Amplitude and Intensity Spatial Interferometry II. Vol. 2200. SPIE; 1994. p. 242. DOI: 10.1117/12.177245

[36] Le Bouquin JB et al. First observations with an H-band integrated optics beam combiner at the VLTI. A&A. 2004;**424**:719-726. DOI: 10.1051/0004-6361:20035849

[37] McAlister HA et al. CHARA Array. In: Breckinridge JB, editor. Amplitude and Intensity Spatial Interferometry II. Vol. 2200. SPIE; 1994. p. 129

[38] Berger JP et al. Integrated optics components for interferometric beam combination. In: Astronomical Interferometry Reasenberg RD, editor. Proceedings of the Astronomical Telescopes and Instrumentation Conference, 20–28 March 1998. Kona, HI, United States. Vol. 3350. SPIE; 1998. p. 898. DOI: 10.1117/12.317158

[39] Malbet F et al. Integrated optics for interferometry. I. Concept and applications Astron. Astrophys. Suppl. Ser. 1999;**138**:1-10. DOI: 10.1051/aas:1999496

[40] Berger JP et al. Integrated optics for astronomical interferometry II. First laboratory white-light interferograms. Astron. Astrophys. Suppl. Ser. 1999;**139**:173-177. DOI: 10.1051/aas:1999504

[41] Haguenauer P et al. Integrated optics for astronomical interferometry III. Optical validation of a planar optics two-telescope beam combiner. Applied Optics. 2000;**39**(13):2130-2139. DOI: 10.1364/AO.39.002130

[42] Berger JP et al. Integrated optics for astronomical interferometry IV. First measurements of stars. A&A. 2001;**376**:L31-L34. DOI: 10.1051/0004-6361:20011035

[43] Laurent E et al. Integrated optics for astronomical interferometry V. Extension to the K band. A&A. 2002;**390**:1171-1176. DOI: 10.1051/0004-6361:20020404

[44] Le Bouquin JB et al. Integrated optics for astronomical interferometry VI. Coupling the light of the VLTI in K band. A&A. 2006;**450**(3):1259-1264. DOI: 10.1051/0004-6361:20054258

[45] Le Bouquin JB et al. First observations with an H-band integrated optics beam combiner at the VLT. A&A. 2004;**424**:719-726. DOI: 10.1051/0004-6361:20035849

[46] Kanaan S et al. Disk and wind evolution of Achernar: The breaking of the fellowship. A&A. 2008;**486**:785-798. DOI: 10.1051/0004-6361:2007886

[47] Bentini GG et al. Integrated Mach-Zehnder microinterferometer on LiNbO3. Optics and Laser Eng. 2007;**45**:368-372. DOI: 10.1016/j.optlaseng.2005.05.006

[48] Bentini GG et al. A new miniaturized optical system for chemical species spectroscopic detection based on a scanning integrated Mach-Zehnder microinterferometer on LiNbO3. Orig. of Life and Ev. of Biosph 2006;**36**:597-603. DOI: 10.1007/s11084-006-9052-3

Chapter 3

2D Relative Phase Reconstruction in Plasma Diagnostics

Michael A. Saville

Abstract

Interferometric analysis methods for measuring plasma properties are presented with emphasis on emerging trends in 2D phase reconstruction. Using recent imagery from exploded-wire experiments the relative phase profiles from independent interferograms are reconstructed. The well-known Fourier Transform Method is presented and discussed. Then, the electron and atom densities are recovered from the phase by solving a linear system of equations in the form of line-integrated density profiles. The mathematical models of the line density and phase function are described and elucidate why interferograms of plasma suffer from low contrast, high signal-to-noise ratio and poorly defined fringes. Although these effects pose challenges for phase reconstruction, the interferometric diagnostic continues to advance the plasma science.

Keywords: Abel transform, density measurement, Fourier transform method, optical interferometry, plasma properties

1. Introduction

This chapter presents an engineering perspective on recent analysis methods used to measure plasma properties from interferograms [1, 2]. The electron and atomic densities [3, 4], volume distribution, expansion velocity, and atomic polarizability [5] affect the amount of fringe line shift. To recover these properties, the basic two-step approach is to recover the relative phase difference represented by the shifted fringe lines, and to invert the relationship between the phase difference and the desired properties.

However, as described in Section 2 and unlike profilometry applications [6] where the interference pattern is stable over a lengthy observation period, the plasma medium changes rapidly and continually. For example, laser ablation methods produce an abruptly expanding plasma lobe. Whereas the electrically exploded wire produces an expanding cylindrical volume that lasts for a few hundred nanoseconds. In each case, the plasma presents an inhomogeneous, lossy and dispersive medium to the probing laser. These medium properties cause radiometric variation in the fringe pattern such as low contrast and poorly defined fringe lines.

Before addressing the fringe line analysis, Section 3 reviews the mathematical model of the phase function for light wave propagation through the plasma volume. The electromagnetic phase accrual through the inhomogeneous medium, and the plasma's refractive index are represented with line integrals of the electron and atom densities. Thus, there are two integral operations to invert during a density measurement. The reader is referred to classic references like [1, 2, 7] for additional factors to consider when the experiment includes controlling magnetic and electric fields. Recent works with electrically exploded wires [8, 9] and dual wavelength interferometry are discussed as a means to recover both density profiles.

Then, Section 4 presents a summary of different fringe analysis methods with emphasis on the Fourier Transform Method (FTM) developed by Takeda, Ina and Kobayashi [10] and continuously improved since [11, 12]. FTM is likely the most well-known method for extracting 2D phase information from the interferogram in surface profilometry and 3D shape measurement. As eloquently described by [13] these applications also give insight into the time–space analog and the time-frequency duality represented by different interferometry experiments.

Section 4 also discusses improvements to FTM which are generally based on iterative filtering [14] or pre-filtering [15]. The latter is presented with examples from exploded wire experiments. Section 5 briefly highlights different works on plasma interferometry and recent advances in the field of plasma science from interferometry. The chapter concludes with some thoughts on how the 2D phase analysis provides a rich understanding of the plasma.

2. Plasma diagnostic setup

2.1 Plasma creation and probing laser setup

Some of the common approaches to create a laboratory plasma include laser ablation, the spark gap, and exploded wires as illustrated in **Figure 1a**. Each drawing shows a

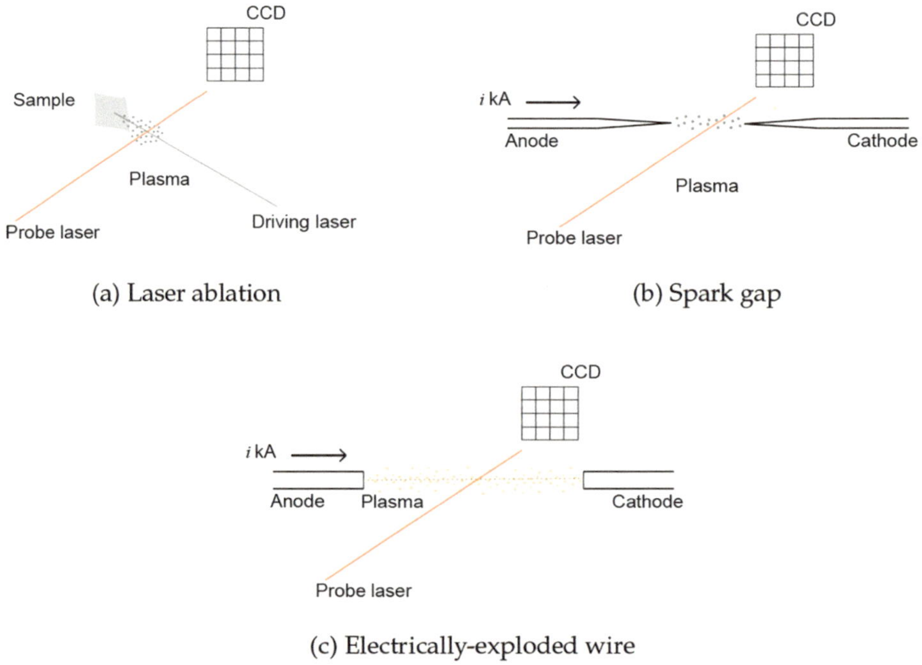

Figure 1.
Illustration of common setup to create plasma and image with a single camera. (a) Driving laser ablates material. (b) Electrical breakdown of air produces electron stream. (c) High driving current vaporizes wire.

probing laser that traverses the plasma medium. Although not shown, the probing laser passes through an interferometer before the charge-coupled-device (CCD) camera captures the interference pattern. **Figure 1a** shows how a driving laser irradiates a sample of very high purity (99.99% or better). The driving laser's power density can range from 10^9 to 10^{12} $\mathrm{W \cdot cm^{-2}}$ for short bursts (≈100 ps) to create a plasma plume or jet [16, 17]. The driving laser's pulse duration and pulse power, and the geometric shape of the sample control the plasma jet's direction and size. The spark gap of [18] as represented in **Figure 1b** uses a pulsed high-voltage to cause the air between the electrodes to break down and form a spark plasma. The low-collisional spark plasma produces a current in air of ≈2.7 kA and lasts for ≈4.2 ms. An alternative approach uses electrically-exploded wires shown in **Figure 1c**. The thin wire samples also of very high purity have very small diameters of a few microns and are driven from the electrodes with a high-current pulse (≈2.5 kA). The pulse has a short rise time of a few nano-seconds and the sudden burst of energy causes the wire to vaporize as a cylindrically expanding plasma [4].

Several recent works use exploding wires to study how the plasma forms and behaves under varying conditions [4, 8, 19]. In particular, the setup of [8] uses two interferograms at 1064 and 532 nm to measure electron and atom densities (illustrated in **Figure 2a**). Also, the addition of a second wire as shown in **Figure 1** allows study of colliding plasma flows [20]. In **Figure 2**, beam splitters and mirrors are understood to change the optical path of probe laser 2.

2.2 Plasma diagnostic methods

Plasma presents electrical, optical and mechanical behaviors that are observable from the electromagnetic (EM) wave emissions in the visible through x-ray regimes and the EM wave propagation through the plasma. Schlieren, shadowgraph and interferometric images reveal temporal and spatial variation of the plasma's index of refraction. Of interest are the interferometer methods such as [8] that measured electron and atom densities using the classical Mach-Zehnder and also the shearing air wedge interferometer [21].

The basic optical setup of [8] as developed from the experiences of [4, 5] is shown in **Figure 3a**. The 1064-nm probing laser is passed through a harmonic doubler to produce a co-linear probing laser at 532 nm. Both beams are adjusted to illuminate and traverse the windowed vacuum chamber. The interferometer is adjusted to provide a regular fringe pattern before wire explosion. **Figure 3b** shows the resulting interference patterns collected at each wavelength (left: 1064 nm, right: 532 nm) at different time intervals before and after wire explosion. The shadowed regions centered near the top and bottom of each image are the electrodes, and the wire's shadow appears in

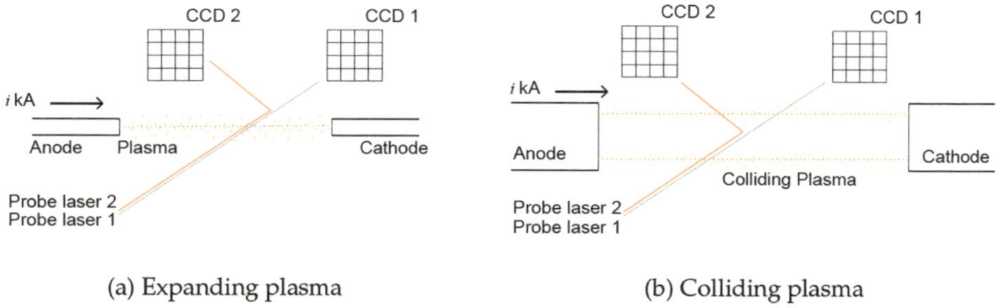

Figure 2.
Illustration of recent methods with electrically exploded wires [8, 20]. (a) Single and (b) double wire setups.

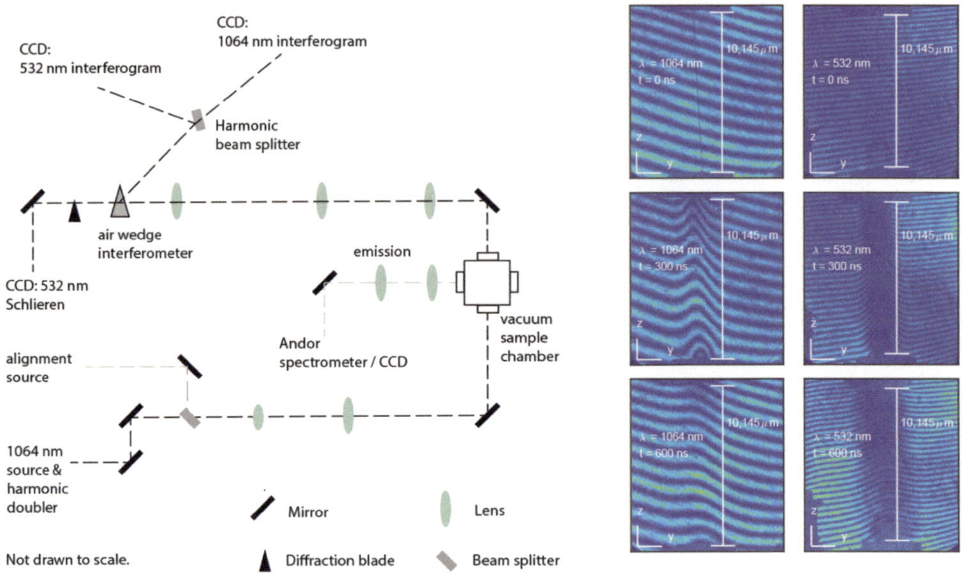

(a) (b)

Figure 3.
(a) Optical path for dual-wavelength interferometry, and (b) time-sequenced interferograms. © 2020 IEEE. Reprinted, with permission, from [8].

the $t = 0$ ns imagery. The CCDs' reference frames are defined with x along the optical path and $y - z$ in the CCD focal planes. Lastly, the setup excludes static electric and magnetic fields, and the laser frequencies are well above the plasma cutoff frequency.

3. Measuring plasma properties with interferometry

3.1 Electron and atom density

In **Figure 4a**, the probe laser passes through the plasma column as a ray and signifies negligible refraction through the inhomogeneous medium. The interferometer (not shown between the plasma and the CCD) establishes the baseline interference pattern and the CCD camera measures the disturbed fringe pattern as caused by the plasma. The shaded discs in the column represent different cross-sectional views of the plasma. Also, the dark to light coloring of **Figure 4b** shows how each particle density has a radial and negative gradient. A more complete model for the density is given in [9]. Referring to **Figure 2**, the Cartesian reference is defined at the junction between the wire and the lower electrode.

In the discussion below, the pixel coordinates (y, z) are described as $\mathbf{r} = \hat{\mathbf{y}}y + \hat{\mathbf{z}}z$, and before wire vaporization ($t = 0$ ns) the total electric field at $\mathbf{r}$ is

$$E(\mathbf{r}) = E_0 e^{ik_0 L}\left[1 + e^{ik_0 \ell} e^{i\Delta\mathbf{k}\cdot\mathbf{r}}\right], \tag{1}$$

where the time convention is $e^{-i\omega t}$ and the model assumes ideal lossless and dispersionless optics for convenience. The common path length before the plasma is L, the length through the plasma region is ℓ, the wave amplitude is E_0, and the

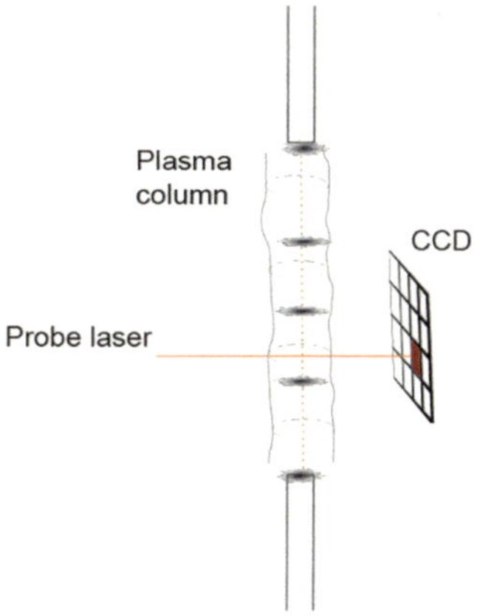

(a) 3D perspective view

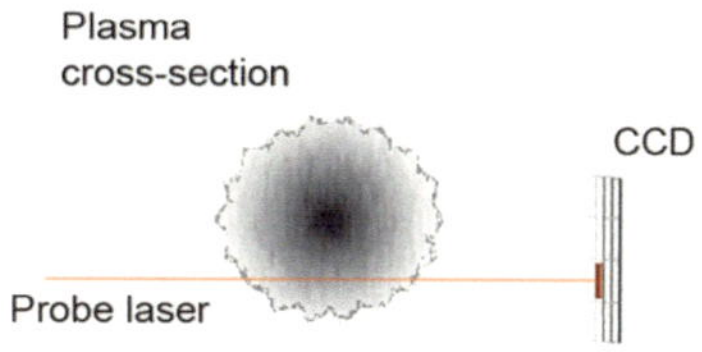

(b) 2D cross-section viewed from top electrode

Figure 4.
Simplified illustration (without refraction) of probe laser passing through plasma of radial density gradient. (a) 3D and (b) 2D vantage of single ray through plasma.

free-space wavenumber is $k_0 = 2\pi/\lambda$ at wavelength λ. In (1), the additional phase term $\Delta\mathbf{k} \cdot \mathbf{r}$ represents the effect of the interferometer which is adjusted to produce straight fringe lines with spatial frequencies $\kappa_y = \Delta\mathbf{k} \cdot \hat{\mathbf{y}}$ and $\kappa_z = \Delta\mathbf{k} \cdot \hat{\mathbf{z}}$. As seen in **Figure 3b** there may be curvature of the fringe lines but those effects are ignored in (1).

Therefore, the intensity pattern in the reference image ($t = 0$ ns) is

$$I_{\text{ref}}(\mathbf{r}) = a_r(\mathbf{r}) + b_r(\mathbf{r}) \cos\left(\Delta\mathbf{k} \cdot \mathbf{r} + k_0 \ell\right), \tag{2}$$

where the background a_r and contrast b_r are proportional to $|E_0|^2$. The non-ideal effects of the optical system cause spatially varying background and contrast [22] that can be measured following the techniques of [23].

After time $t = 0$ ns the plasma has wavenumber $k = k_0 n$ and the propagation phase through the plasma is $k_0 \int_\ell n \mathrm{d}x$ [24]. The test image has intensity

$$I_{\text{tst}}(\mathbf{r}) = a_t(\mathbf{r}) + b_t(\mathbf{r}) \cos\left[\Delta\mathbf{k} \cdot \mathbf{r} + k_0 \int_\ell n(x, \mathbf{r}) \mathrm{d}x\right], \tag{3}$$

where the index of refraction n is inhomogeneous and the background and contrast are proportional to $|E_0|^2 / \int_\ell n \mathrm{d}x$.

The index of refraction, assuming complete ionization is expressed with the atomic and electron densities N_a and N_e, respectively, as

$$n(x, \mathbf{r}) = 1 + \alpha N_a(x, \mathbf{r}) - \beta N_e(x, \mathbf{r}). \tag{4}$$

where α is the dynamic polarizability of the material, $\beta = \beta_0 \lambda^2$, $\beta_0 = e^2 / \left(16\pi^3 \varepsilon_0 m_e c_0^2\right)$, ε_0 is the free-space permittivity, m_e is the electron mass, c_0 is speed of light in vacuum, and e is unit charge.

Upon substituting (4) into (3) it is clear how the fringe spacing in the interferogram is determined by $\Delta\mathbf{k} \cdot \mathbf{r}$ and the amount of line shift is determined by the line-integrated densities

$$\chi_a(\mathbf{r}) = \int_\ell N_a(x,\mathbf{r})\,dx, \tag{5}$$

$$\chi_e(\mathbf{r}) = \int_\ell N_e(x,\mathbf{r})\,dx. \tag{6}$$

When the plasma has axial symmetry and radial density profiles as illustrated in **Figure 4**, Eqs (5) and (6) can be expressed in terms of the forward Abel transform $\mathcal{A}\{\cdot\}$ as [25].

$$\chi_a(y,z) = \mathcal{A}\{N_a\} = 2\int_y^\infty \frac{N_a(r,z)y}{\sqrt{r^2-y^2}}\,dr, \tag{7}$$

$$\chi_e(y,z) = \mathcal{A}\{N_e\} = 2\int_y^\infty \frac{N_e(r,z)y}{\sqrt{r^2-y^2}}\,dr. \tag{8}$$

where $r^2 = x^2 + y^2$. The upper limit of the integration is usually restricted to the radius R of the plasma column.

Upon expanding the phase argument of (3) with (4)–(6) as

$$\Phi(y,z) = \kappa_y y + \kappa_z z + k_0\ell + k_0\alpha\chi_a(y,z) - k_0\beta\chi_e(y,z), \tag{9}$$

and comparing with the phase argument of (2) the phase difference caused by the plasma is

$$\Delta\phi(y,z) = k_0\alpha\chi_a(y,z) - k_0\beta\chi_e(y,z). \tag{10}$$

In cases where $N_a \ll N_e$, $\Delta\phi \approx -k_0\beta\chi_e$ and N_e is measured using the inverse Abel transform as

$$N_e(r,z) = \mathcal{A}^{-1}\left\{\frac{-\Delta\phi(y,z)}{k_0\beta}\right\} = -\frac{1}{\pi}\int_r^\infty \frac{\partial\chi_e(y,z)}{\partial y}\frac{1}{\sqrt{y^2-r^2}}\,dy. \tag{11}$$

Otherwise, two independent wavelength measurements may be used to solve for the line-integrated densities. Then, each radial density profile is determined with Abel inversion of the respective line-integrated density [4]. Also, when the amount of line shift is normalized by the spacing between undisturbed fringe lines the resulting fringe ratio is equivalent to the normalized phase difference $\delta = \Delta\phi/(2\pi)$. Therefore, δ has implicit units of lines. When the interferogram has high contrast or high signal-to-noise ratio (SNR) it is straightforward to visually calculate δ and then apply the inverse Abel transform.

The time-sequenced, 1D fringe ratios of [8] are shown in **Figure 5** and were calculated by visual inspection of the average fringe shift in each image. A contour-fitting algorithm was also proposed to calculate $\delta(y)$, but the 2D calculation of $\delta(y,z)$ with FTM is presented in Section 4.3.

3.2 Expansion velocity

The expansion velocity is calculated from the change in the plasma's radial dimension as observed over time. Schlieren imagery and shadowograms are better suited to

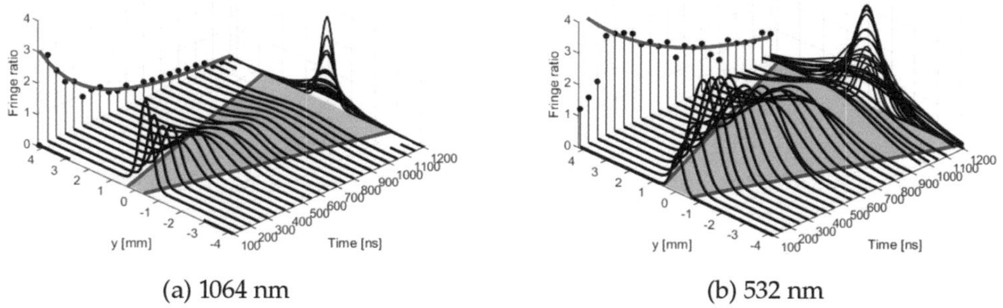

Figure 5.
Time-sequenced fringe ratio from applying a contour-fitting algorithm to Cu interferograms. © 2020 IEEE. Reprinted, with permission, from [8].

visually observe the precise change of the plasma radius over time, but the changes are observable in the interferogram. The process is to measure the span ΔD_y where the fringe line begins to shift away from the reference line and where it just returns to the reference line. A common approach is to determine D_y from each image in the region where $\delta > 0.1$ [26]. With D_{y1} and D_{y2} equal to the spans at times t_1 and t_2, the expansion velocity is $(D_{y2} - D_{y1})/(t_2 - t_1)$ as shown by the shaded regions of **Figure 4**. The accuracy is limited by the pixel size and thickness and clarity of the fringe line. Also, the expansion speed is highly dependent on the method used to produce the plasma. Speeds of a few km/s are commonly observed in wire experiments but simulation of colliding flows shows speeds at high orders of Mach [20].

3.3 Ionization ratio

The ionization ratio N_e/N_a is highly dependent on the experimental conditions. For example, excess electrons from the excitation current must be considered. Also, some plasmas will have greater concentration of different types of ions. Therefore, the refractivity model should include terms for the higher order ions as in [1]. It has been noted that it is unlikely one can measure individual ion density from a single interferogram. In particular, the work of [8] should be considered as an effort to measure electron and atom densities rather than electron and ion densities.

3.4 Polarizability

Atomic polarizability is measurable from interferometry as described in [5]. The authors of [5] use the integrated-phase technique that equates the density of a cross-sectional slice of the wire with the corresponding integrated phase inferred by the interferogram. With the linear density defined as $N_{\text{lin}} = \iint N_a \mathrm{d}x\mathrm{d}y$ the dynamic polarizability is measured as

$$\alpha(\lambda) = \frac{k_0}{N_{\text{lin}}} \int_{D_y} \chi_a(y)\, \mathrm{d}y, \tag{12}$$

where the z dependency is suppressed, and D_y is the span of plasma according to the where the fringe shifts occur. The authors of [5] also carefully note the assumptions of total vaporization, negligible free-electron refraction, and that the shifted fringe lines represent the region of the plasma. Under the conditions they report dynamic polarizability with an accuracy of 10% for Mg, Ag, Al, Cu, and Au samples at laser wavelengths of 1064 and 532 nm. The measured values of Al as given in [5] are used in the example of Section 4.6.

4. Relative phase reconstruction with Fourier transform method (FTM)

4.1 Relative and absolute phase reconstruction

Before addressing the analysis methods, it is helpful to note how the measured phase is relative to a known reference. Unlike radio frequency methods, the phase reference is generally unknown in optical interferometry. Thus, the term phase reconstruction implicitly means *relative* phase reconstruction. The phase difference developed in Section 3.1 is the subject of the phase reconstruction methods and referred to as the phase function and phase profile in this section.

4.2 1D fringe analysis

When the interference pattern has a uniform behavior along one dimension such as $\kappa_y = 0$ then each fringe line can be visually analyzed to measure the amount of line shift. In the case of a uniformly tilted line, i.e. $\kappa_y \neq 0$ and $\kappa_z \neq 0$ then the image can be rotated such that $\kappa_y = 0$. Also, the principle component analysis can be performed on the fringe line to measure the shift, but firstly, requires a method to extract the fringe line. The approach in [8] treated the intensity image as a surface and extracted the contours for each fringe line in the image. After averaging, normalization and fitting of the lines to a Gaussian curve for each wavelength, the line-integrated densities were recovered and inverted with the Abel transform to calculate the electron and atom densities. The 1D approach is unlikely to work well for closed loop fringe lines that occur in surface profilometry and collisional plasma interferometry.

4.3 2D fringe analysis

The 2D fringe analysis is applicable to temporal and spatial analysis, 1D and 2D interference patterns, and straight lines and open and closed curves. Approaches to reconstruct $\phi(y, z)$ include principal component analysis [27, 28], phase shifting [29], wavelet analysis [30] and Fourier analysis [10, 15]. The latter has become popular in profilometry, 3D shape reconstruction and more recently measurement of nano-scale and femtosecond observables. In much of the literature the original work of [10] has become known as the Fourier Transform Method (FTM).

FTM is well-known for 2D phase reconstruction in optical metrology of surface flatness [31], surface height, strain [32] and defect [33] profilometry, surface motion [34], and 3D shape measurement. However, the spatial-carrier fringe behavior has also been used to observe extreme physical phenomenon [35] such as magnetic fields, electron waves and ultra-violet lithography, as well as beam propagation [36], plasma property measurement [37], refractive index studies of polymeric substrates [38], nano-scale surface metrology [39], corneal topography [40] and biological tissue characterization [41].

The essence of the approach is to analyze the 2D Fourier spectrum of the fringe pattern and to extract the phase function using a digital homodyne receiver process. In treating the interferogram as a 2D signal with spatial dimensions and spatial frequencies κ_y and κ_z, the perturbation is a modulating signal to be recovered as in a communication system. FTM first demodulates the image by the spatial frequency and then filters the baseband content. The solution is succinctly and elegantly described as a two-step algorithm [6, 12, 13, 35]. The listing in Algorithm 1 includes two additional steps to estimate the spatial frequencies and to unwrap the phase function.

Algorithm 1. Fourier Transform Method [12].

Given input image $I(y,z) = a(y,z) + b(y,z)\cos\left[\kappa_y y + \kappa_z z + \phi(y,z)\right]$.
Measure spatial frequencies κ_y and κ_z.
Demodulate I to baseband and low-pass filter using the Fourier Transform as

$$c(y,z) = h(y,z) * \left[I(y,z)e^{i\kappa_y y}e^{i\kappa_z z}\right], \tag{13}$$

Recover the wrapped phase ϕ as

$$\phi(y,z) = \arctan\left(\frac{\mathcal{Im}\{c(y,z)\}}{\mathcal{Re}\{c(y,z)\}}\right). \tag{14}$$

Unwrap phase: $\tilde{\phi} = unwrap(\phi)$.
return unwrapped phase $\tilde{\phi}(y,z)$.

In (13), h is an ideal 2D rectangular window function with size $Q_y \times Q_z$, and $Q_y = \lceil 2\pi/\kappa_y \rceil$ and $Q_z = \lceil 2\pi/\kappa_z \rceil$ where $\lceil \cdot \rceil$ denotes the ceiling function. The last step to unwrap the phase may be accomplished with a variety of techniques. To demonstrate the FTM approach, an interference image is simulated as $I(y,z) = a(y,z) + b(y,z)\cos\left[\kappa_y y + \kappa_z z + \phi(y,z)\right]$ with spatial frequencies $\kappa_y = \pi/5 \text{ rad} \cdot \text{m}^{-1}$ and $\kappa_z = \pi \text{ rad} \cdot \text{m}^{-1}$. The phase function ϕ is defined as

$$\phi(y,z) = \pi \exp\left[-0.1(z/10 + y - 0.1)^2\right], \tag{15}$$

and the background and contrast functions are specified as

$$a(y,z) = 0.6 + \gamma\delta(y,z), \tag{16}$$

$$b(y,z) = 1. \tag{17}$$

In (16), $\gamma \sim \mathcal{N}(0, \sigma^2)$ is a random variable from the standard normal distribution with zero mean and variance σ^2. The variance is set to produce phase noise with 10-dB signal-to-noise ratio.

The simulated phase and interferogram are shown in **Figure 6a** and **b**. The phase profile causes fringe lines similar to what are observed with a single exploded wire. However, the radiometric effects are ignored in the example. The results of the different stages of Algorithm 1 are also shown in **Figure 6c–f**. The spectrum of g has three components: a direct-current (DC)-like term, the spatial carrier, and the carrier's conjugate.

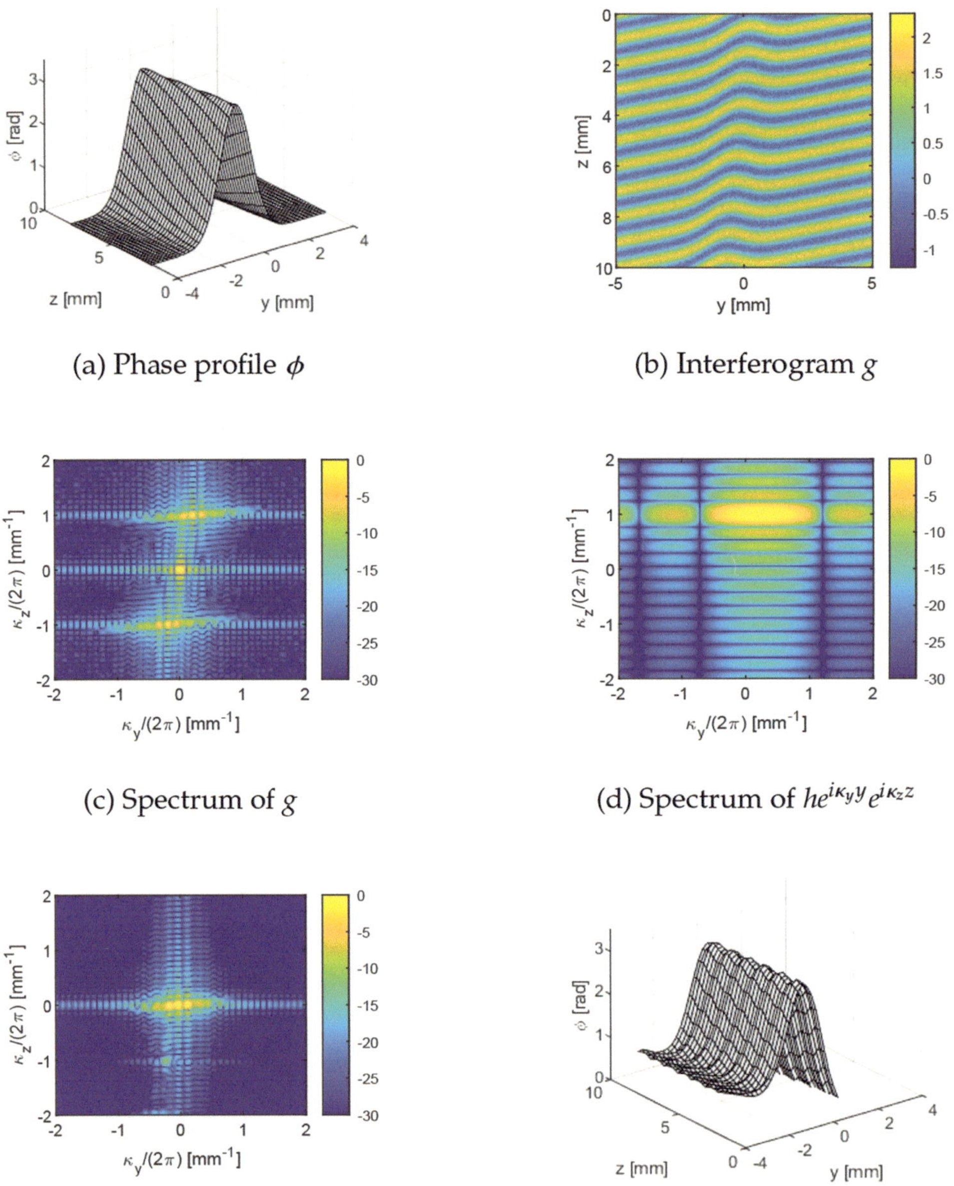

(a) Phase profile ϕ

(b) Interferogram g

(c) Spectrum of g

(d) Spectrum of $he^{i\kappa_y y}e^{i\kappa_z z}$

(e) Spectrum of c, i.e. g after filtered demodulation

(f) Unwrapped reconstructed phase $\tilde{\phi}$

Figure 6.
Example of FTM using simulated noisy 532-nm interferogram. (a) phase, (b) image, (c) signal spectrum, (d), filter spectrum, (e) filtered signal spectrum, (f) recovered phase.

The DC component centered at $\kappa_y = \kappa_z = 0$ is mainly due to the background a but often includes residual energy from the carrier and its conjugate. The region of interest is centered at the spatial carrier frequency (κ_y, κ_z) and has a spectral shape determined by $c = be^{i\phi}$. Therefore, the phase recovery is sensitive to the design of the filter h. **Figure 6f** shows the result when the ideal rectangular impulse response is used per [12] and has a noticeable ripple and effect of Gibbs phenomenon. The root mean square error between ϕ and the reconstructed and unwrapped phase $\tilde{\phi}$ is 0.67 radians.

4.4 FTM with plasma interferometry

Figure 7 shows examples of measured plasma interferograms (left column). The images exhibit significant intensity variation and regions of very low contrast. The top and middle images have 10–15 dB SNR and the bottom image has 5 dB SNR. The spatial frequency spectra are shown in the middle column with a bounding box drawn

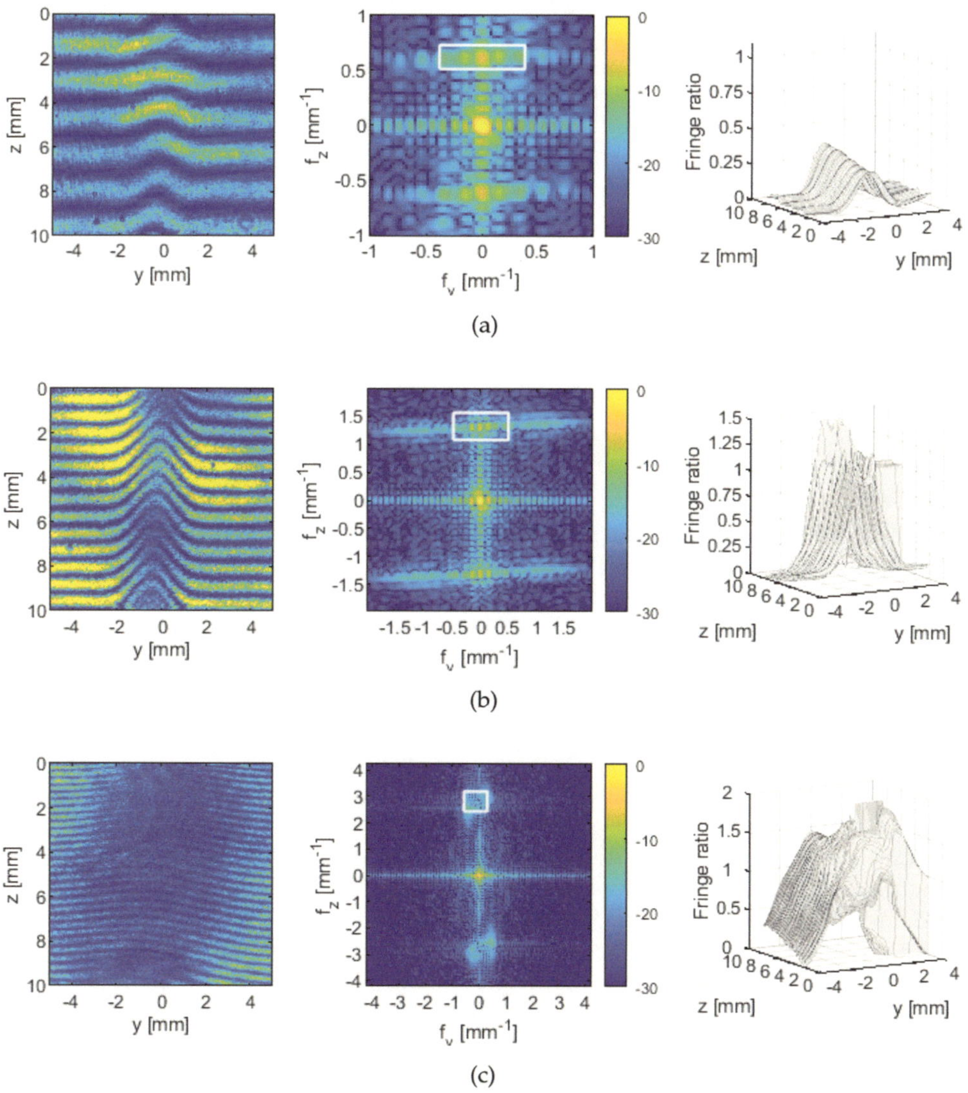

Figure 7.
Interferograms (left column), spatial frequency spectrum (middle column) with filter domain shown as a black box, and reconstructed fringe order (right column). The spatial frequencies are $f_y = \kappa_y/(2\pi)$ and $f_z = \kappa_z/(2\pi)$. Examples with (a) low spatial frequency, (b) low SNR, and (c) low contrast and low SNR.

about the positive carrier frequency. The box has dimensions of an ideal 10-dB bandwidth rectangular convolution filter. In the top image, the energy located about the upper carrier frequency is spread and overlaps with the DC content in the center of the image. Lastly, the fringe lines in some of the images show high frequency ripple caused by the interferometer.

From the frequency images in the middle column of **Figure** 7 it is easy to see how the FTM filter h requires careful design [42–45]. However, there are various tradeoffs between accuracy, computation time, and filter design. Such details are often left out of the discussion on FTM's role in phase reconstruction. The effects in the imagery affect the accuracy of the phase reconstruction and ultimately the electron and atom density measurements. The right column of **Figure** 7 shows the fringe ratio as recovered with FTM. In the top row, the vertical sidelobes of the DC energy bleeds through the bandpass filter at the carrier causing ripple in the fringe order. Note: the DC content is shifted to the carrier in FTM, i.e. Eq. (13) of Algorithm 1. The intensity variation of the middle row is due to intensity variation of the probing laser and the inhomogeneous plasma volume. The spatial dependencies of a and b cause additional frequency spreading about the carrier. Likewise, the bottom interference pattern has very low contrast, i.e. low SNR and the energy about the carrier is very weak. Thus, the energy spectra are distorted and the fringe ratios are coarsely recovered as shown in **Figure 7b** and **c**.

4.5 Improvements to FTM

Given the various image artifacts that hinder phase reconstruction, several FTM improvements have been developed to improve the phase accuracy of interference patterns with high density fringe lines. The different approaches can be considered as pre-filtering or iterative filtering of $I(y,z) = a(y,z) + b(y,z)\cos\left(\kappa_y y + \kappa_z z + \phi_S(y,z)\right)$. The most prominent is the iterative model-based approach of [14] that uses Zernike polynomials to model the phase profile ϕ_S and then iteratively improve the model using narrowing filters. In context of **Figure** 7, the algorithm of [14] first finds the initial model ϕ_{Mdl} using conventional FTM and removes the background a with a DC rejection filter. However, some residual energy from the DC region remains causing the ripple observed in reconstructed phase ϕ_S and as observed in the fringe order of **Figure** 7. The phase function is fitted with a suitable 2D smooth polynomial which effectively excludes the rippling. Using the modeled phase function $\phi_{S,\mathrm{Mdl}}$, the original demodulated and DC-rejected image is iteratively filtered to remove the conjugate phase. The extracted phase error $\Delta\phi = \phi_S - \phi_{S,\mathrm{Mdl}}$ is used to update the model until the phase ϕ_S converges. The authors of [14] report only three iterations were needed in their simulations and the phase accuracy improves by a factor of ten.

The pre-filtering approach proposed in [8] filters the background function a and artificially equalizes the envelope function as $b(y,z) = 1$. The equalization algorithm is given in [46] and the approach is shown by simulation to improve phase reconstruction by a factor of five. While the accuracy is less than the iterative method, it is much simpler to implement. However, it is limited to open fringe lines and is unknown to perform as well as the iterative method for high-density fringe lines. **Figure 8** shows examples of the smoothing and leveling algorithm of [46] with the measured interferograms of **Figure** 7. The fringe lines are noticeably improved and easy to analyze by visual inspection. Also, the spectra show a significant reduction of DC content. The resulting fringe ratio plots have less ripple and the density profiles shown in **Figure 8b**

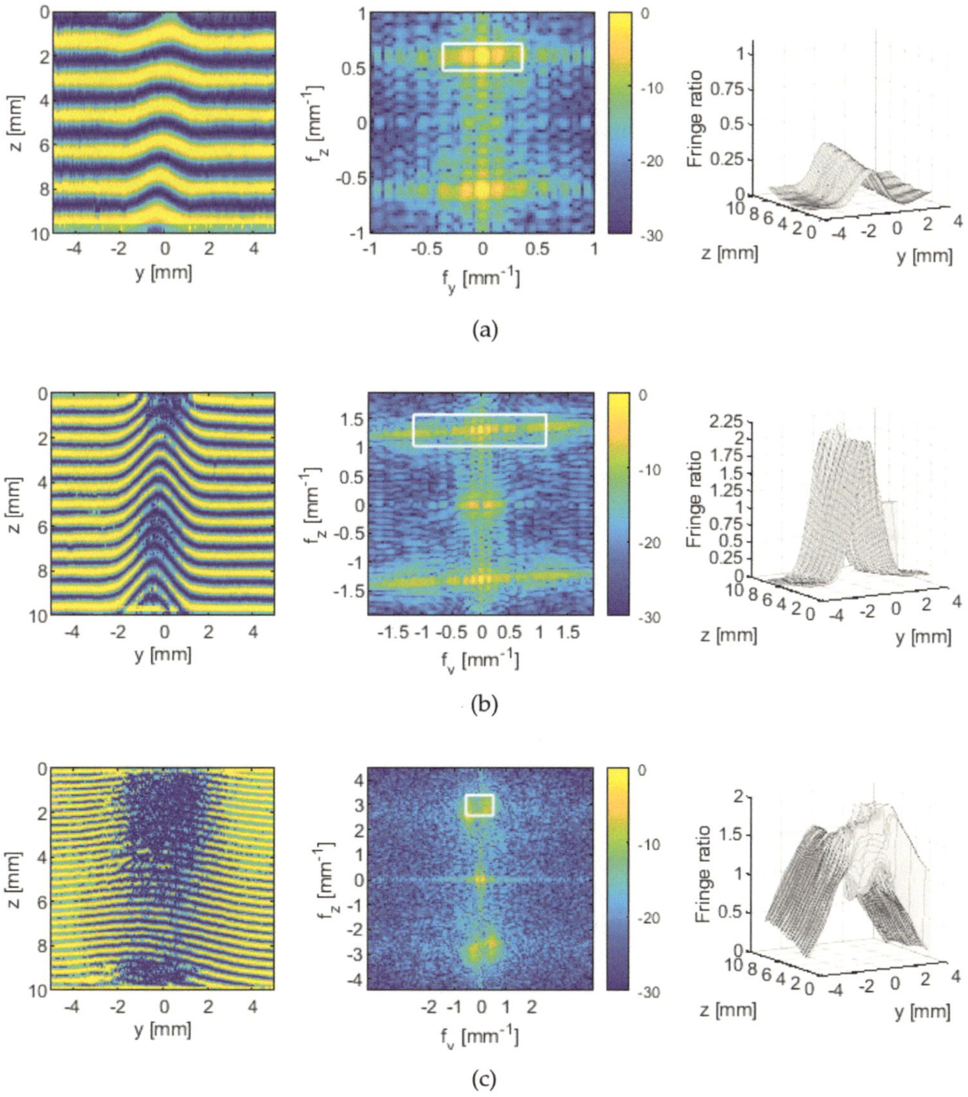

Figure 8.
*Smoothed and leveled interferograms of **Figure 6** (left column), spatial frequency spectrum (middle column) with filter domain shown as a black box, and FTM reconstructed fringe order (right column). Recovered from image with (a) low spatial frequency, (b) low SNR, and (c) low contrast and low SNR.*

and **c** are improved over **Figure 7b** and **c**. In context of the iterative FTM algorithm [14], the phase accuracy using the pre-filtering approach is limited because the residual energy of the signal conjugate can still bleed into the final baseband signal. However, the pre-filtering approach could serve as the initial phase estimate.

4.6 Example of electron and atomic density measurement

To demonstrate the density measurement the procedure of [8] is adapted for 2D electron and atomic density measurements. The pre-filtered interference images are

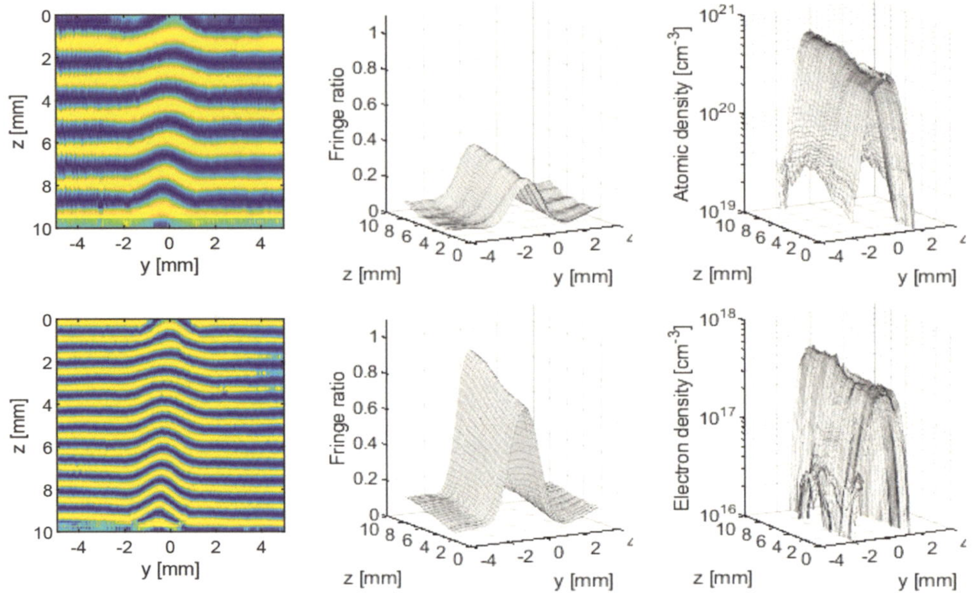

(a) (b) (c)

Figure 9.
2D phase reconstruction and density measurement of Al plasma following set up of [8]. (a) 1064-nm (top) and 532-nm (bottom) interference patterns. (b) 1064-nm (top) and 532-nm (bottom) fringe ratios, and (c) atomic (top) and electron (bottom) volumetric densities (cm^{-3}).

shown in **Figure 9a** and the measurement setup and wire dimensions are given in [8]. Each phase profile is reconstructed with FTM and presented as fringe ratio (**Figure 9b**). From the two independent line shift measurements $\delta_1 = \Delta\phi_1/(2\pi)$ (1064 nm) and $\delta_2 = \Delta_2/(2\pi)$ (532 nm), the line-integrated densities χ_a and χ_e are determined from the linear system of equations

$$\delta = \mathrm{A}\chi, \tag{18}$$

where $\boldsymbol{\delta} = [\delta_1, \delta_2]^T, \boldsymbol{\chi} = [\chi_a, \chi_e]^T$, the superscript T denotes the transpose operator, and the matrix

$$\mathbf{A} = \begin{bmatrix} \alpha_1/\lambda_1 & -\beta_0\lambda_1 \\ \alpha_2/\lambda_2 & -\beta_0\lambda_2 \end{bmatrix}, \tag{19}$$

is easily inverted. In (19), the dynamic atomic polarizabilities at 1064 and 532 nm are $\alpha_1 = 8.7 \times 10^{-24}$ and $\alpha_2 = 10.8 \times 10^{-24}$ cm^3, respectively [5]. The explicit expressions for the line-integrated densities are

$$\chi_a = \Delta(-\beta_0\lambda_2\delta_1 + \beta_0\lambda_1\delta_2), \tag{20}$$

$$\chi_e = \Delta(-\alpha_2\delta_1/\lambda_2 + \alpha_1\delta_2/\lambda_1), \tag{21}$$

where $\Delta = \lambda_1\lambda_2/\left(\alpha_2\beta_0\lambda_1^2 - \alpha_1\beta_0\lambda_2^2\right)$. The line-integrated densities are transformed into the volumetric densities (**Figure 9c**) by applying the inverse Abel Transform as

$$\begin{Bmatrix} N_a(r,z) \\ N_e(r,z) \end{Bmatrix} = \mathcal{A}^{-1} \begin{Bmatrix} \chi_a(y,z) \\ \chi_e(y,z) \end{Bmatrix}, \tag{22}$$

4.7 Measurement accuracy

The accuracy of the density measurement depends on the accuracies of (1) the line shift measurement, and (2) the numerical accuracy of the Abel inversion. The numerical accuracy of the transform is determined by the specific implementation. Therefore, the direct integration methods [25] depend on the pixel size but can easily achieve double precision accuracy. Thus, the primary source of measurement error is due to the relative phase shift (i.e., line shift). A visual analysis of the fringe pattern has the best accuracy because the line shift can be measured to ± 1 pixel. The relative error is $\pm 100/N_m\%$ where N_m is the number of pixels at the maximum line shift. However, for the generic 2D pattern or when the image has noise and speckle, the accuracy depends on the complete phase reconstruction method. According to [15], the overall phase measurement algorithm includes pre-filtering to reduce noise and speckle, relative phase recovery such as with FTM, and phase unwrapping, and the phase evaluation should be better than $\pi/10$ rad for general purposes. The error analysis of [5], also showed the phase-shift reconstruction accuracy should be on the order of $2\pi/20$ which is equivalent to 0.05 lines. For a maximum line shift δ_M, the accuracy is $\pm 100/(20\delta_M)\%$. For an interferogram with up to 3 or 4 lines of shift, the accuracy is estimated as $\pm 7\%$. Other approaches to show improvement of the phase measurement first determine the phase difference with the phase-shifting method [27], and then compare it to each of the errors from the iterative FTM [14] and the conventional FTM. The error bound is given by the peak-to-valley measurement and the average error is given as root-mean-square error. The improved FTM with iterative filter narrowing is shown to improve accuracy by a factor of ten.

An open problem is to perform a rigorous sensitivity analysis. It can be completed with modeling and simulation using the following model:

$$\begin{bmatrix} \hat{N}_a \\ \hat{N}_e \end{bmatrix} = \mathcal{A}^{-1} \left\{ \mathbf{A}^{-1} \mathcal{F} \left\{ \begin{bmatrix} a + b\cos\left(\Delta\mathbf{k}_1 \cdot \mathbf{r} + g(\mathcal{A}\{N_a\}, \mathcal{A}\{N_e\})\right) \\ a + b\cos\left(\Delta\mathbf{k}_2 \cdot \mathbf{r} + g(\mathcal{A}\{N_a\}, \mathcal{A}\{N_e\})\right) \end{bmatrix} \right\} \right\}, \tag{23}$$

where g is the mapping of the simulated line-integrated densities to the phase function described in (9) of Section 3. The measured densities are denoted as $\hat{N}_a$ and $\hat{N}_e$, and $\mathcal{F}$ denotes the FTM or other phase reconstruction (line shift measurement) method.

By appropriate choice of noise terms affecting a and b (e.g., multiplicative phase noise, additive Gaussian noise, fringe line distortions, radiometric variation or other optical aberrations), (23) can determine the effects on accuracy due to different parameters in the phase recovery algorithm and measurement setup.

5. Advances in plasma diagnostics with interferometry

There are numerous examples of plasma diagnostics with optical and x-ray interferometry. This section highlights a few to illustrate the specific advances in plasma science. Much of the interferometry science with plasma seeks to reconstruct the signal phase and refractive index of the plasma as a means to measure the electron density. However, the various types of plasmas and interferometers have shown additional benefits of the interferometric diagnostic.

For example, Feister et al. [47] developed a temporally multi-scale interferometer to investigate the target pre-ablation by ultra intense pulses ($>10^{18}$ W $\cdot$ cm^{-2}). The setup acquired three-phase interference images using a laterally sheared Michelson setup. The combination of phase reconstruction from the interferogram and the ultra-fast shadowgraphy confirmed how self-emission that corrupted the electron density measurement was caused by pre-ablation of the sample. The femtosecond time scale confirmed that the leading edge of the intense pulse caused the pre-ablation and led to images showing nanosecond formation of pre-plasma, femtosecond interaction of the ultra-intense main pulse, and picosecond hydrodynamic expansion. The physical insights gained by the different time-sequenced views were applied to modeling and simulation and subsequent hardware design. In the x-ray approach of Nilsen and Johnson [48], a 14.7-nm Pd laser x-ray interferometry confirmed how bound electrons caused anomalous fringe behavior in Al plasma that resulted in a refractive index less than one.

Since then, other works [9, 20] investigated physical behavior of expanding plasma volumes when created from two electrically exploded wires. In [9], aluminum (Al) and polymide-coated tungsten (W) plasma from a 1-kA current was observed with dual-wavelength (532 and 1064 nm) interferometry. They found that the atomization of Al expanded with a constant velocity of several km/s before stagnating in the middle region of the two wires. They concluded the Al plasma comprised mostly atoms as observed by the density calculations and comparison to the linear wire density. They also discussed how the W wires were more difficult to transition to vapor due to the higher melting point and concluded that a dual-pulse generator would be needed in future experiments. A similar study of mixing flow of two expanding plasma was reported in [20] where they also used two electrically exploded wires to study the hot plasma coronas as they collided. The modeling and simulation study showed that a thin layer between the expanding coronas can be sustained in high Mach number flows resulting from hydrodynamic mixing. It also showed how the dense core material enhances a thin shell instability but has yet to be mapped to observation.

In the recent work of [23], the spontaneous magnetic fields that arise during laser ablation were studied using simultaneous measurement of the polarization plane rotation and plasma electron density. A 10^{16} W $\cdot$ m^{-2} iodine laser irradiated planar and thick Cu targets to generate a plasma in air. However, unlike other types of plasma studies, the phase ϕ and amplitude b are necessary to measure the polarization state. Thus, the paper presents a detailed summary of FTM and intermediate steps to recover the amplitude. The introduction of [23] includes a historical perspective of the polaro-interferometry methods for SMF measurement. Ultimately, the work reported a new multi-frame complex interferometry system that can measure the distributions of the magnetic field and electron density from the same interferogram.

6. Conclusions

Advances in plasma science greatly benefit from optical and x-ray interferometry. The first part of the chapter summarized the plasma experiment setup and the models for the plasma refractivity and the interferogram's 2D signal phase. The Abel inversion technique was described for cases where the electron density greatly exceeds the atom density. However, the optical interferometer can cause significant radiometric variation in the image, low contrast, and low signal-to-noise. These artifacts require special attention for accurate phase reconstruction. Of the various techniques to recover the phase from a single image (e.g, principle component analysis, contour fitting of the fringe lines, and Fourier analysis), the Fourier Transform Method (FTM) is presented with detailed algorithmic steps. Additionally, recent iterative improvements to FTM and a simpler smoothing and leveling pre-filtering algorithm are highlighted. While it is less capable than the iterative method, the pre-filtering approach is demonstrated using dual-wavelength interferometry of exploded Aluminum wires.

However the plasma is formed, the interference patterns are well-suited for 2D Fourier analysis and several plasma experiments were highlighted. The experiments have confirmed previously theorized observations about the effects of bound electrons on refractive index and how precisely the plasma forms during ablation, and how it expands, emits, and eventually recombines. As bandpass and polarimetric filters and beamsplitters improve, and cameras increase in acquisition speed and sensitivity, the interferometric diagnostic will also improve to provide new and greater understanding of the plasma.

Acknowledgements

The author sincerely thanks Andrew Hamilton and Dr. Vladamir Sotnikov for support of previous work and helpful discussions, and Michael Gruesbeck and Javonne Baker for testing the smoothing and leveling algorithm. Lastly, this work was greatly improved by the review of Dr. Gennady Sarkisov whose vast experience with exploding wire experiments and plasma diagnostics motivated the initial work.

Conflict of interest

The author declares no conflict of interest.

Thanks

Thank you Xiomy for your steadfast love and faith as we live to honor Him.

Author details

Michael A. Saville
Wright State University, Dayton, Ohio, USA

*Address all correspondence to: michael.saville@wright.edu

References

[1] Huddlestone R, Leonard S, editors. Plasma Diagnostic Techniques. New York, San Francisco and London: Academic Press; 1965

[2] Lovberg RH, Griem HR, editors. Plasma Physics Part B (Methods of Experimental Physics). New York and London: Academic Press; 1971

[3] Pikuz SA, Romanova VM, Baryshnikov NV, Hu M, Kusse BR, Sinars DB, et al. A simple air wedge shearing interferometer for studying exploding wires. Review of Scientific Instruments. 2011;**72**(1):1098-1100

[4] Sarkisov GS, Rosenthal SE, Cochrane KR, Struve KW, Deeny C, McDaniel DH. Nanosecond electrical explosion of thin aluminum wires in a vacuum: Experimental and computational investigations. Physical Review E. 2005;**71**(046404):1-21

[5] Sarkisov G, Beigman I, Shevelko V, Struve K. Interferometric measurements of dynamic polarizabilities for metal atoms using electrically exploding wires in a vacuum. Physical Review A. 2006; **73**:042501

[6] Takeda M, Yamamoto H. Fourier-transform speckle profilometry: Three-dimensional shape measurements of diffuse objects with large height steps and/or spatially isolated surfaces. Applied Optics. 1994;**33**(34):7829-7837

[7] Chen FF. Introduction to Plasma Physics and Controlled Fusion. New York and London: Plenum Press; 1984

[8] Hamilton A, Saville MA, Sotnikov V. Dual-wavelength interferometric measurement of electrically exploded aluminum and copper wires in low-pressure air for electron and ion density calculation. IEEE Transactions on Plasma Science. 2020;**48**(9):3144-3151

[9] Wu J, Li X, Lu Y, Lebedev SV, Yang Z, Jia S, et al. Atomization and merging of two Al and W wires driven by a 1 kA, 10 ns current pulse. Physics of Plasmas. 2016; **23**(112703)

[10] Takeda M, Ina H, Kobayashi S. Fourier-transform method of fringe-pattern analysis for computer-based topography and interferometry. Journal of the Optical Society of America. 1982; **72**(1):156-160

[11] Takeda M. Temporal versus spatial carrier techniques for heterodyne interferometery. In: Arsenault HH, editor. Optics and the Information Age, International Society for Optics and Photonics. Vol. 0813. SPIE; 1987. pp. 329-330

[12] Takeda M. Spatial-carrier fringe-pattern analysis and its applications to precision interferometry and profilometry: An overview. Industrial Metrology. 1990;**1**(2):77-99

[13] Takeda M. Optical metrology: Methodological analogy and duality revisited. In: Mahajan VN, Kim D, editors. Tribute to James C. Wyant: The Extraordinaire in Optical Metrology and Optics Education, International Society for Optics and Photonics. Vol. 11813. SPIE; 2021. pp. 134-141

[14] Nakayama S, Toba H, Fujiwara N, Gemma T, Takeda M. Enhanced Fourier-transform method for high-density fringe analysis by iterative spectrum narrowing. AIP Advances. 2020;**59**(29): 9159-9164

[15] Hipp M, Woisetschläger J, Reiterer P, Neger T. Digital evaluation of

interferograms. Measurement. 2004; **36**(1):53-66

[16] Grava J, Purvis MA, Filevich J, Marconi MC, Dunn J, Moon SJ, et al. Soft X-ray laser interferometry of a dense plasma jet. IEEE Transactions on Plasma Science. 2008;**36**(4):1286-1287

[17] Doyle LA, Martin GW, Al-Khateeb A, Weaver I, Riley D, Lamb MJ, et al. Electron number density measurements in magnesium laser produced plumes. Applied Surface Science. 1998;**127–129** (716):720

[18] Patra AS, Phukan TD, Khare A. Measurement of two-dimensional electron density profile in a low current spark using interferometry. IEEE Transactions on Plasma Science. 2005; **33**(5):1725-1728

[19] Han R, Li C, Yuan W, Ouyang J, Wu J, Wang Y, et al. Experiments on plasma dynamics of electrical wire explosion in air. High Voltage. 2022;**7**(1): 117-136

[20] Sotnikov VI, Hamilton A, Malkov MA. Collision of expanding plasma clouds: Mixing, flow morphology, and instabilities. Physics of Plasmas. 2020;**27**(12):122113

[21] Sarkisov G. Shearing interferometer with an air wedge for the electron density diagnostics in a dense plasma. Instruments and Experimental Techniques. 1996;**09**(39):727-731

[22] Sarkisov GS. Anomalous transparency at 1064 nm of a freely expanding gas cylinder in vacuum during fast electric explosion of thin metal wires. Journal of Applied Physics. 2022;**131**(10):105904

[23] Zaraś-Szydłowska A, Pisarczyk T, Chodukowski T, Rusiniak Z, Dudzak R, Dostal J, et al. Implementation of amplitude–phase analysis of complex interferograms for measurement of spontaneous magnetic fields in laser generated plasma. AIP Advances. 2020;**10**(11): 1-20

[24] Fitzpatrick R. Plasma Physics an Introduction. Boca Raton, London and New York: CRC Press; 2015

[25] Hickstein DD, Gibson ST, Yurchak R, Das DD, Ryazanov M. A direct comparison of high-speed methods for the numerical Abel transform. Review of Scientific Instruments. 2019;**90**(6):065115

[26] Smith RF, Moon S, Dunn J, Nilsen J, Shlyaptsev VN, Hunter JR, et al. Interferometric diagnosis of two-dimensional plasma expansion. AIP Conference Proceedings. 2002;**641**(1): 538

[27] Fan J, Lu X, Xu X, Zhong L. Simultaneous multi-wavelength phase-shifting interferometry based on principal component analysis with a color CMOS. Journal of Optics. 2016; **18**(5):055703

[28] Vargas J, Sorzano COS, Estrada JC, Carazo JM. Generalization of the principal component analysis algorithm for interferometry. Optics Communications. 2013;**286**:130-134

[29] Dubois A. A simplified algorithm for digital fringe analysis in two-wave interferometry with sinusoidal phase modulation. Optics Communications. 2017;**391**:128-134

[30] Zhong J, Weng J. Spatial carrier-fringe pattern analysis by means of wavelet transform: Wavelet transform profilometry. Applied Optics. 2004; **43**(26):4993-4998

[31] Maaboud N, El-Bahrawi M, Abdel-Aziz F. Digital holography in flatness and crack investigation. Metrology and Measurement Systems. 2010;**XVII**(4): 583-588

[32] Sur F, Blaysat B, Grédiac M. Determining displacement and strain maps immune from aliasing effect with the grid method. Optics and Lasers in Engineering. 2016;**86**: 317-328

[33] Hu E, Zhu Y, Shao Y. Defect information detection of a spare part by using a dual-frequency line-scan method. Optik. 2014;**125**(3):1255-1258

[34] Rivera-Ortega U, Pico-Gonzalez B. Computational tool for phase-shift calculation in an interference pattern by fringe displacements based on a skeletonized image. European Journal of Physics. 2016;**37**(015303):1-9

[35] Takeda M. Fourier fringe analysis and its application to metrology of extreme physical phenomena: A review invited. Applied Optics. Jan 2013;**52**(1): 20-29

[36] Popiolek-Masajada A, Borwinska M, Przerwa-Tetmajer T, Kurzynowski P. Application of the Fourier analysis methods to the three beam interferometry. Optics Laser Technology. 2013;**06**(48):503-508

[37] Zhang H, Lu J, Ni X. Optical interferometric analysis of colliding laser produced air plasmas. Journal of Applied Physics. 2009;**106**(6):063308-1-063308-5

[38] Singh G, Mehta DS. Measurement of change in refractive index in polymeric flexible substrates using wide field interferometry and digital fringe analysis. Applied Optics. 2012;**51**(35): 8413-8422

[39] Ye S, Takahashi S, Michihata M, Takamasu K, Hansen H, Calaon M. Quantitative depth evaluation of microgrooves on polymer material beyond the diffraction limit. Precision Engineering. 2019;**59**:56-65

[40] Micali JD, Greivenkamp JE. Dual interferometer for dynamic measurement of corneal topography. Journal of Biomedical Optics. 2016; **21**(8):85007

[41] van Werkhoven TIM, Antonello J, Truong HH, Verhaegen M, Gerritsen HC, Keller CU. Snapshot coherence-gated direct wavefront sensing for multi-photon microscopy. Optics Express. 2014;**22**(8):9715-9733

[42] Huntley JM. Automated fringe pattern analysis in experimental mechanics: A review. The Journal of Strain Analysis for Engineering Design. 1998;**33**(2):105-125

[43] Jesus C, Arasa J, Royo S, Ares M. Design of adaptive digital filters for phase extraction in complex fringe patterns obtained using the Ronchi test. Journal of Modern Optics. 2012;**04**(59): 721-728

[44] Zhao B, Asundi AK. Criteria for phase reconstruction using Fourier transformation method. In: Fiddy MA, Millane RP, editors. Image Reconstruction from Incomplete Data, International Society for Optics and Photonics. Vol. 4123. SPIE; 2000. pp. 269-278

[45] Kujawińska M. The architecture of a multipurpose fringe pattern analysis system. Optics and Lasers in Engineering. 1993;**19**(4):261-268

[46] Saville MA. Unpublished Notes on Fourier Transform-Based Phase

Reconstruction with Smoothed and Leveled Imagery; 2022

[47] Feister S, Nees JA, Morrison JT, Frische KD, Orban C, Chowdhury EA, et al. A novel femtosecond-gated, high-resolution, frequency-shifted shearing interferometry technique for probing pre-plasma expansion in ultra-intense laser experiments. Review of Scientific Instruments. 2014;**85**(11):11D602

[48] Nilsen J, Johnson WR. Plasma interferometry and how the bound-electron contribution can bend fringes in unexpected ways. Applied Optics. Dec 2005;**44**(34):7295-7301

Chapter 4

Bragg Grating Tuning Techniques for Interferometry Applications

Rogério Dionísio

Abstract

Fiber Bragg grating is widely used in optical fiber applications as a filter or a sensor due to its compact size and high sensitivity to physical conditions, such as temperature and strain. The purpose of this chapter is to describe the implementation and characterization of two tuning methods for optical fiber Bragg gratings, varying the temperature or the length of the fiber. Among the methods using mechanical deformation, compression of the fiber by bending a flexible sheet aggregated with the Bragg grating has shown very interesting tuning results, reaching 19.0 nm with minimum reflection bandwidth variation over the entire tuning range. Stretching the fiber has presented several drawbacks, including breaking of the fiber and a lower tuning range of 4.9 nm. Temperature tuning technique presents good linearity between tuning range and temperature variation but at the cost of a low tuning range (0.4 nm) and a permanent high current electrical source.

Keywords: fiber Bragg grating, tuning system, thermal tuning, mechanical tuning, experimental characterization

1. Introduction

Fiber optic Bragg gratings (FBGs) were first produced in 1978 by Ken Hill et al., during an experiment where an optical fiber was exposed to intense radiation from an argon laser. The interference process started between the incident wave and the 4% light reflected at the opposite end of the fiber [1]. It was not until a decade later, in 1989, that G. Meltz and his team proposed Bragg gratings production process by exposing the fiber transversely to two coherent beams of ultraviolet light, forming an interference pattern in the fiber core [2, 3].

Over the last decades, Bragg gratings have been used as sensors [4, 5] or as optical filters in wavelength division multiplexing (WDM) communication systems [6]. They have found application in devices with routing capacity in the optical domain, as is the case of optical add drop multiplexers (OADM) and optical cross connect (OXC) nodes. They are also used in frequency stabilizers of semiconductor lasers [7], on tunable fiber lasers [8], and in the gain equalization of optical amplifiers [9]. Fiber dispersion compensation [10] is another realization with tunable fiber Bragg gratings.

The evolution of transmission systems from point-to-point connections to dynamically configurable networks brought the need for tunable optical filters. Furthermore, the increasing transfer of information over fiber optic networks causes an increase in the number of optical channels in Dense (DWDM) or even in Coarse WDM (CWDM). The expansion in the bandwidth necessity range can cover the S, C, L, and U bands, corresponding approximately to 215 nm.

During the last 3 decades, but also very recently, the use of spread spectrum techniques in the optical domain, such as optical code division multiplexing (OCDMA), has stirred the interest of the scientific community, as an option in reconfigurable optical networks [11]. This process allows the various channels to share the same spectral band; each one being identified by a specific code. Other features of this technique should be highlighted: Greater safety in the data transmission, more flexible use of available bandwidth, reducing crosstalk between adjacent channels, and enabling asynchronous communication [12]. The recent use of spectral amplitude coding OCDMA (SAC-OCDMA) in optical coding and decoding systems has brought the ability to eliminate multiple access interference (MAI) [13]. For this technique to be attractive from the point of view of implementation in an optical network, the definition of the codes will have to be flexible [14], so that tunable FBGs are a key element of the process. One of the features that make FBGs attractive is that the reflection spectrum can be tuned from a few nanometers, heating the Bragg grating or applying mechanical tension to the ends (compression or extension).

The objective of this chapter is to describe the implementation of two tuning methods for fiber Bragg gratings, as a function of variations in temperature or in the length of the FBG. Among the methods of mechanical deformation, the deformation of the grating by stretching or compressing the fiber and the deformation by curvature of a flexible blade connected to the Bragg grating are experimentally implemented and discussed.

The rest of the chapter is structured as follows: Section 2 presents the operating principle of FBG tuning techniques using mechanical and temperature techniques. Section 3 describes the experimental measurement setup for FBG tuning characterization. Section 4 presents the main experimental results and discusses the potential of each tuning technique. Finally, Section 5 gives concluding remarks.

2. Operating principle

The resonant wavelength (λ_B) of a Bragg grating depends on the effective refractive index at the fiber core (n_{eff}) and the period of the interference pattern (Λ) [15],

$$\lambda_B = 2 \bullet n_{eff} \bullet \Lambda \tag{1}$$

In turn, these parameters are affected by temperature variations or mechanical deformations. From **Eq. (1)**, the deviation of the Bragg wavelength due to mechanical deformations Δl, or temperature variations ΔT [16] are:

$$\Delta\lambda_B = 2\left(\Lambda\frac{\delta n_{eff}}{\delta l} + n_{eff}\frac{\delta\Lambda}{\delta l}\right)\Delta l + 2\left(\Lambda\frac{\delta n_{eff}}{\delta T} + n_{eff}\frac{\delta\Lambda}{\delta T}\right)\Delta T \tag{2}$$

The first term of **Eq. (2)** represents the dependence of λ_B as a function of Δl, caused by variations in the grating period and the change in the effective refractive index. This term can be described by:

$$\Delta\lambda_B = (1 - \rho_e) \bullet \varepsilon_{rel} \bullet \lambda_B, \quad (3)$$

where ε_{rel} is the relative elongation and ρ_e is the effective elasto-optic coefficient. For a silica optical fiber, $\rho_e \approx 0.22$ [17].

The second term of expression **(2)** represents the effect of temperature on changing the Bragg wavelength. For a Bragg grating with maximum reflectivity at 1550 nm, the typical values of sensitivity to temperature and mechanical stress are, respectively, 13.25 pm/°C and 1.2 pm/με [16].

It is therefore predictable that any change in wavelength associated with external disturbances is the result of the sum of the effects of temperature and mechanical deformations. Therefore, the individual discrimination in the spectral response of the FBG for each source of disturbance needs a method of separating the measurements. Briefly, the existing methods are extrinsic or intrinsic and are reported in detail in [16]. In Section 4.1, a set of tuning experiments is described, based on stretching or compressing the fiber, and which take place in a laboratory in a controlled temperature environment. Under these conditions, the influence of temperature on the wavelength variation is very small compared to the variations induced by the mechanical compression or stretching methods.

2.1 Mechanical linear tuning

Figure 1 illustrates the elements that make up a tuning system by mechanical tension on the fiber. The ends of the optical fiber are fixed on two supports, where at least one must be able to move along the axial axis of the fiber.

The relative elongation along the axial axis is defined by

$$\varepsilon_{rel} = \frac{\Delta z}{L}, L \neq 0, \quad (4)$$

where Δz is the displacement and L is the length of the fiber under the effect of a mechanical deformation [18]. Substituting this result into **Eq. (3)**, we get:

$$\Delta\lambda_B = (1 - \rho_e) \bullet \frac{\Delta z}{L} \bullet \lambda_B, L \neq 0 \quad (5)$$

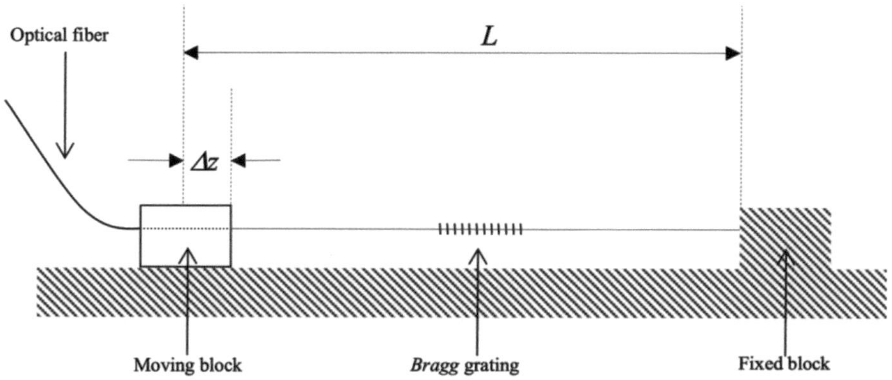

Figure 1.
Schematic of the structure of a tuning system for fiber optic Bragg networks.

The result is an equation that relates the variation of the Bragg wavelength as a function of the normalized fiber length. The change in wavelength will be positive or negative if the Bragg grating undergoes stretching or compression, respectively. Considering that the displacement is much smaller than the length of the grating, **Eq. (5)** is approximately linear.

2.2 Mechanical tuning by bending

Figure 2 shows the scheme of a tuning system, in which a force is applied to a flexible blade. Horizontal displacement Δz causes the blade to bend in an arc with angle θ. The relationship between the angle arc θ and the displacement Δz is given by:

$$\Delta z = L\left[1 - \frac{sin\left(\frac{\theta}{2}\right)}{\left(\frac{\theta}{2}\right)}\right], \tag{6}$$

where L is the length of the blade in the initial state, with no force applied to it [19, 20].

For an optical fiber embedded in an elastic material at a distance d from the blade, the relative elongation is given by [19–21]:

$$\varepsilon_{rel} = \mp\frac{d \bullet \theta}{L}, L \neq 0 \tag{7}$$

The negative sign corresponds to a compressive force when the blade is bent upwards, while a positive sign in the second term of **Eq. (7)** indicates a pulling force on the fiber, bending the blade downwards.

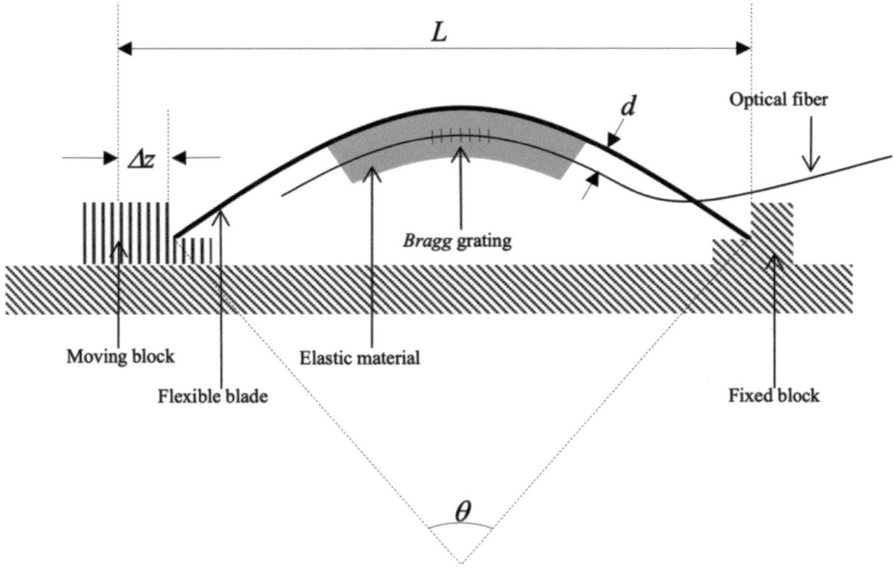

Figure 2.
Schematic of the structure of a tuning system based on bending a flexible blade.

Conjugating **(3)**, **(6)**, and **(7)** into a single equation,

$$\frac{\Delta z}{L} = 1 - sinc\left(\frac{\Delta\lambda_B \bullet L}{2 \bullet \pi \bullet d \bullet (1-\rho_e) \bullet \lambda_B}\right), L \neq 0 \tag{8}$$

The relationship between the spectral tuning λ_B and the displacement is not linear, contrary to the result in the procedure in Section 4.1.

2.3 Temperature tuning

The second term of **Eq. (2)** provides an explicit relationship of the dependence of the Bragg wavelength on temperature variations and is expressed by:

$$\Delta\lambda_B = (\alpha_T + \xi_T) \bullet \Delta T \bullet \lambda_B, \tag{9}$$

where α_T is the thermal expansion coefficient and ξ_T is the thermo-optical coefficient. In the case of silica, the mentioned constants have the following values: $\alpha_T = 0.55 \times 10^{-6}K^{-1}$ and $\xi_T = 8.0 \times 10^{-6}K^{-1}$.

A possible tuning scheme is to place the Bragg grating over a thermoelectric module, also called a Peltier cell, and envelop the set-in thermal mass. The objective is to standardize the temperature in the network and increase the thermal conductivity between the optical fiber and the Peltier module [22, 23].

It is also possible to increase the thermal sensitivity of the optical fiber by fixing it on a metallic surface. If a zinc (Zn) sheet is used, it is possible to increase the sensitivity to 40.8 pm/K. By fixing two Peltier modules at the ends of the zinc sheet, a temperature gradient is created that causes a linear variation of the grating period, which is used to tune the Bragg wavelength, maintaining a constant group delay. For this, the temperature difference between the two ends is kept constant, while the temperature at both ends is increased or decreased by the same amount [24].

3. Laboratory measurement system

To characterize the tuning methods, an optical measurement system was set up, whose implementation scheme is shown in **Figure 3**. It is based on the measurement of the spectral component reflected by the network when illuminated by an optical

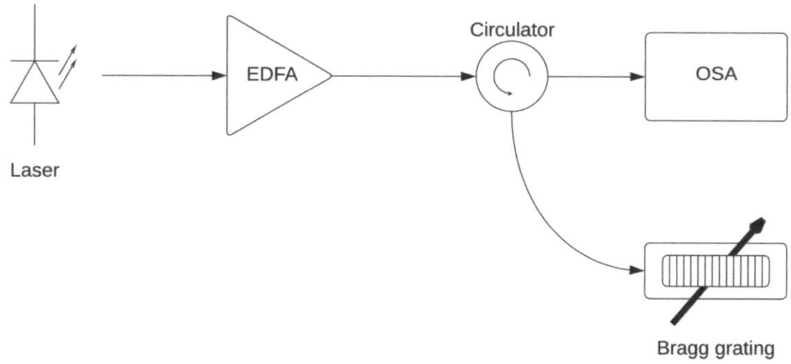

Figure 3.
Experimental setup used to determine the deviations in the central wavelength of the FBG, caused by the tuning system.

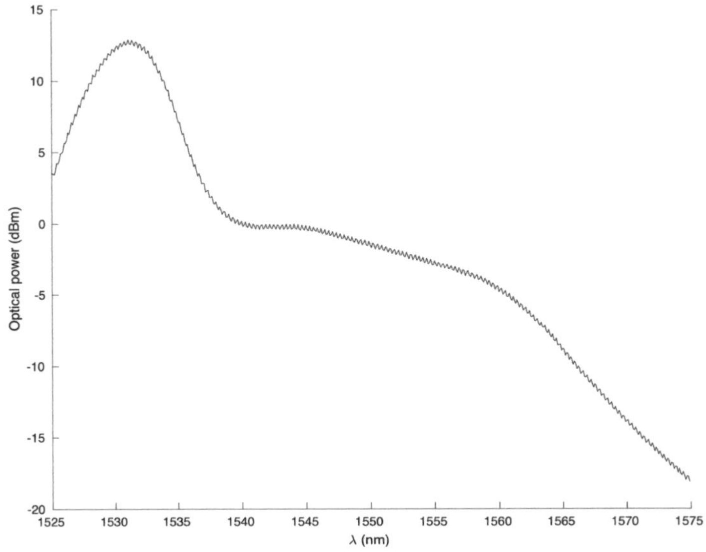

Figure 4.
Optical spectrum of the ASE measured at the amplifier output for an optical pumping power value of 88 mW.

source with high spectral bandwidth. The input optical signal is amplified and introduced into the Bragg grating through an optical circulator. The signal reflected by the FBG is then conveyed by the circulator to an optical spectrum analyzer (ANRITSU model MS9601A). The Bragg grating is tuned by a motorized system developed for this purpose, which is described in ref. [25].

A commercial optical amplifier (Photonetics model BT 13) with a pumping laser at 980 nm and an optical saturation power of 13 dBm was used.

Figure 4 shows the ASE spectrum, measured with a resolution of 0.5 nm, where the dependence of the spectral amplitude with the wavelength is clearly visible. Therefore, all spectral measurements described in the following chapter are represented by subtracting the optical power of the source from the reflection spectrum, at each wavelength.

Bragg gratings are recorded on photosensitive single-mode fiber (Fibercore model PS1250/1500) with numerical aperture 0.13 and diameter 125 μm. The etching method exposes the fiber to ultraviolet radiation through a constant period phase mask. The recording optical source is an Argon laser operating in continuous mode at 244 nm and with an average power of 150 mW. A more detailed description of the recording system can be found in [26, 27]. The apodization format of the network is approximately Gaussian because the laser used has an optical beam with a Gaussian profile.

4. Experimental results and discussion

4.1 Mechanical tuning

The tuning test, based on the scheme in **Figure 1**, is applied to a Bragg network with 1.5 cm long, half-height bandwidth 0.1 nm, center wavelength at 1556.94 nm, and rejection range of 10 dB. The length of the fiber containing the FBG is 18 cm, corresponding to the initial distance between the clamping presses.

According to studies carried out by several authors [18, 19], the maximum relative elongation that can be exerted on a silica optical fiber is 1%. To preserve the elasticity characteristics of the fiber, a more conservative value, around 0.5%, is considered the maximum limit. Regarding the fiber length used, it corresponds approximately to 0.9 mm. In the computerized tuning process, described in [25], this limit is considered to prevent the optical fiber from breaking.

Figure 5 shows the reflection spectrum of the stretched Bragg grating. The maximum deviation from the central wavelength is 4.9 nm and corresponds to an offset of 0.98 mm. According to **Eq. (3)**, the relative elongation was 0.4%, remaining within the preestablished limit.

Figure 6 shows the evolution of the bandwidth at 1, 3, and 10 dB, during the stretching process of the Bragg grating. Specifically, the bandwidth at 1 dB exhibits a variation within the range $\pm$ 0.01 nm, for the bandwidth of 3 dB there is an increase of 0.01 nm, and at 10 dB there is an increase of 0.05 nm in the bandwidth. Changes in maximum reflectivity are less than 1.1 dB.

The relationship between the variation of the central wavelength and the relative elongation is shown in **Figure 7**. The deviation from the central wavelength grows linearly as a function of relative elongation, at a rate of 0.85 pm/με and with a correlation coefficient of 0.9951. The expected value for the growth rate, using expression **(3)**, would be 1.2 pm/με.

To verify whether this discrepancy is caused by imperfections in the motorized system, the influence of positioning errors on central wavelength tuning is calculated:

$$|d(\Delta\lambda_B)| \leq \left|\frac{\partial(\Delta\lambda_B)}{\partial(\Delta z)}\right| |d(\Delta z)| = \frac{0.78\lambda_B}{L} |d(\Delta z)|, \tag{10}$$

where $|d(\Delta z)|$ is the tuning system accuracy (15 μm), L is the fiber length (18 cm) and λ_B is the Bragg wavelength of the Bragg grating at rest (1556.94 nm). With these

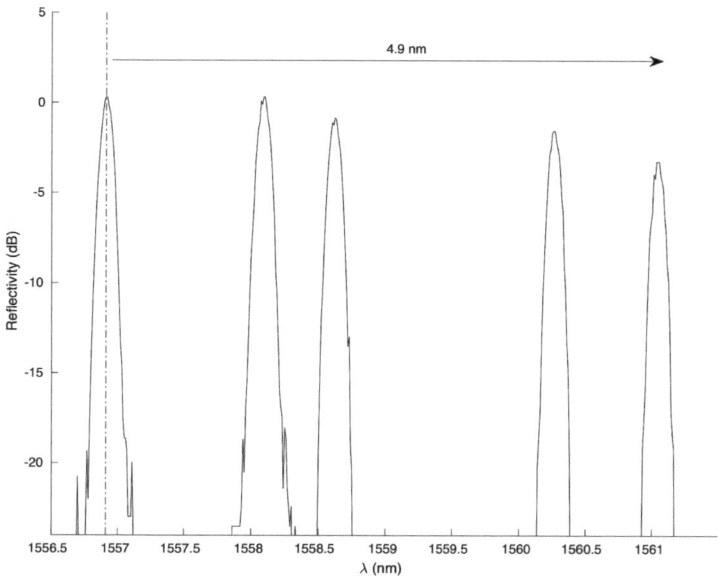

Figure 5.
Reflection profile of a Bragg grating under the effect of mechanical strain.

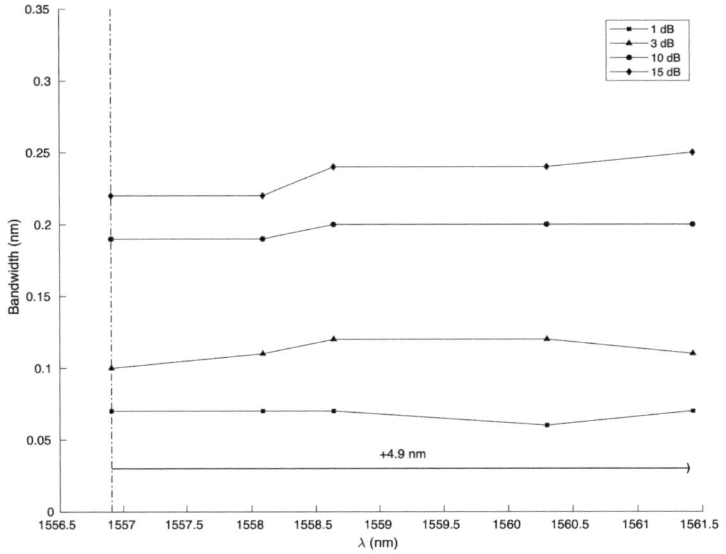

Figure 6.
Bandwidth evolution over the entire tuning range.

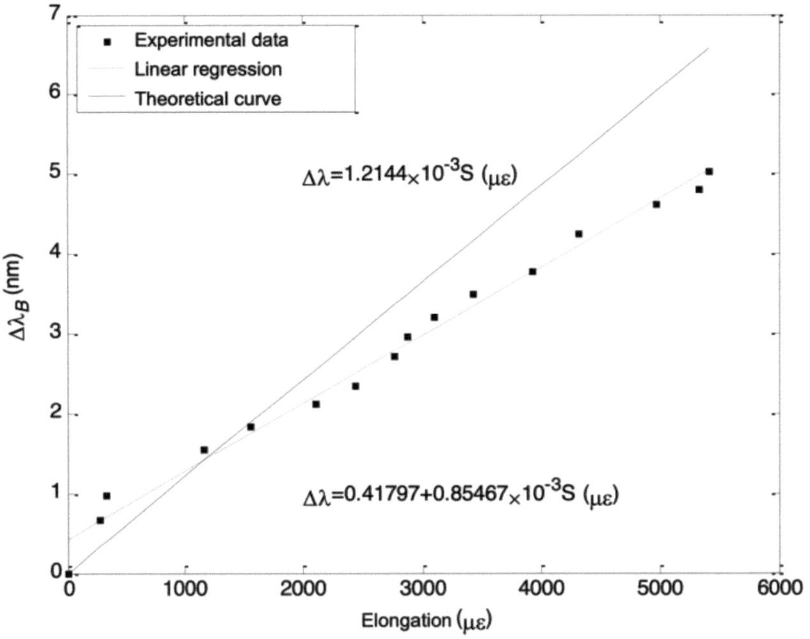

Figure 7.
Bragg wavelength variation as a function of relative elongation.

values, the variation of the tuning value $|d(\Delta\lambda_B)|$ is always less than or equal to 101 pm. If the error $|d(\Delta\lambda_B)|$ is included in the set of measurements performed for the central wavelength, the resulting growth rate is in the range [0.84228, 0.85467] pm/με. Therefore, imperfections in the motorized system are not the main cause of the differences.

A second tuning test had the scheme described in Section 2.2 as its working principle. In the first stage, the fiber is not embedded in an elastic material but fixed directly to the surface of a flexible sheet. The blade used is 15.2 cm long, 2 cm wide, and 1.5 mm thick, and is made of acrylic. A "v" shaped groove is created along the blade, 9 mm deep, to accommodate the fiber, so that it does not suffer transverse deviations. Consequently, d = 6 mm (see **Eq. (7)**). The assembly is then glued to increase its strength and to properly fix the FBG inside the groove. Finally, the support is placed between the moving and fixed blocks of the tuning system to start the tests.

It was said earlier that the fiber can be stretched up to 1% without suffering irreversible damage. In compression mode, this value can reach 23% [19], which allows much larger tuning ranges than in elongation. Substituting **Eq. (7)** in **(6)**, the resulting equation allows calculating a priori the maximum displacement allowed without damaging the fiber. Using more conservative values of 0.5% for compression and 11% for relative elongation, the calculated limits are, respectively, 9.96 mm and 14.13 cm.

The Bragg grating used is 2 cm long, it has a 3 dB bandwidth equal to 0.48 nm and a central wavelength of 1550.4 nm. During the tests, the amplitude of the reflection spectrum was recorded by an optical spectral analyzer, at several tuning points, as shown in **Figure 8**.

The results show that the variation achieved in the central wavelength, using compression, and stretching forces reaches practically 19 nm. It is visible in **Figure 8** that there is a gap in the initial phase of the stretching tuning process. This failure occurred simply because the spectral measurement process was started with the Bragg grating already under the effect of a sufficiently strong mechanical stress that the central wavelength was shifted from its initial value of 1550.4 nm.

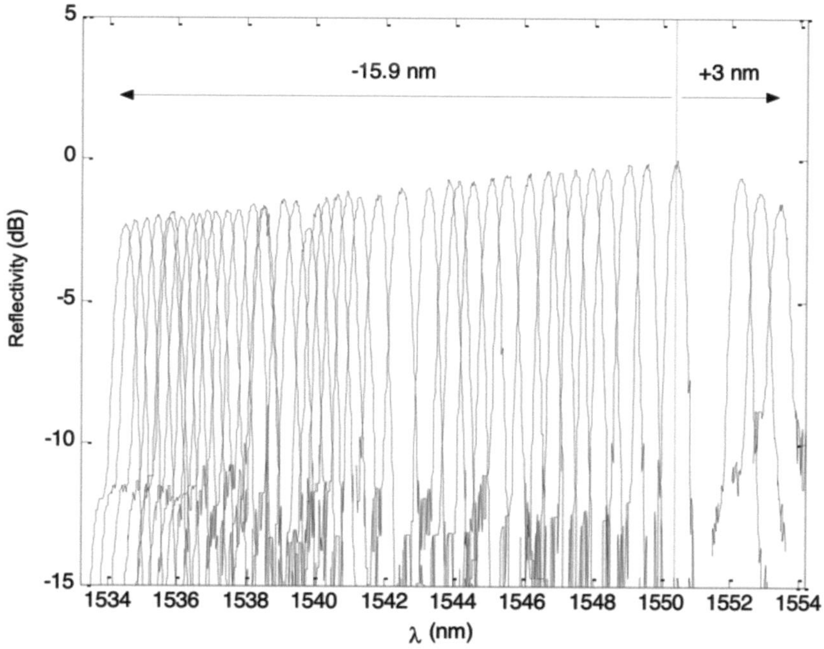

Figure 8.
Tuning the central wavelength of the Bragg grating at 18.9 nm.

It is possible to obtain better results, especially in compression mode, as the mobile platform moved only 6.8 cm to obtain a variation of −15.9 nm. However, the mechanical characteristics of the acrylic sheet did not allow to compress the fiber further. On the other hand, the 3 nm spectral variation in stretching mode was achieved by moving the system 4.3 mm and could theoretically reach 4.2 nm with a displacement equal to 9.96 mm.

Figure 9 shows the relationship between tuning range $\Delta\lambda$ and normalized length $\Delta z/L$.

The results show a good agreement between the theoretical curve, calculated by **Eq. (8)**, and most of the measurements performed.

The main differences are located at the ends of the graph, which correspond to situations of greater mechanical stress. They are somehow correlated with the progressive decrease in the maximum reflectivity amplitude, visible in **Figure 8**. These changes can be caused by undetected problems in the tuning system. In particular, the glue used can lose its fixing qualities when subjected to large deformations, resulting in bends on the fiber [18].

The evolution of the bandwidth at 1, 3, and 10 dB is represented in **Figure 10**. Starting from the initial position, the axial mechanical forces applied on the FBG do not cause great changes in the width at 3 dB, which oscillates between −0.1 and 0.04 nm. The bandwidth at 1 dB also remains practically constant, with a maximum deviation of 12.5% from the initial value. However, the bandwidth at 10 dB undergoes a gradual increase that reaches 0.4 nm in compression and 0.56 nm in stretching, in addition to showing an irregular behavior.

The increase in bandwidth at 10 dB is clearly visible in **Figure 11**, where the reflection spectrum is not symmetrical, especially when the FBG is subjected to strong

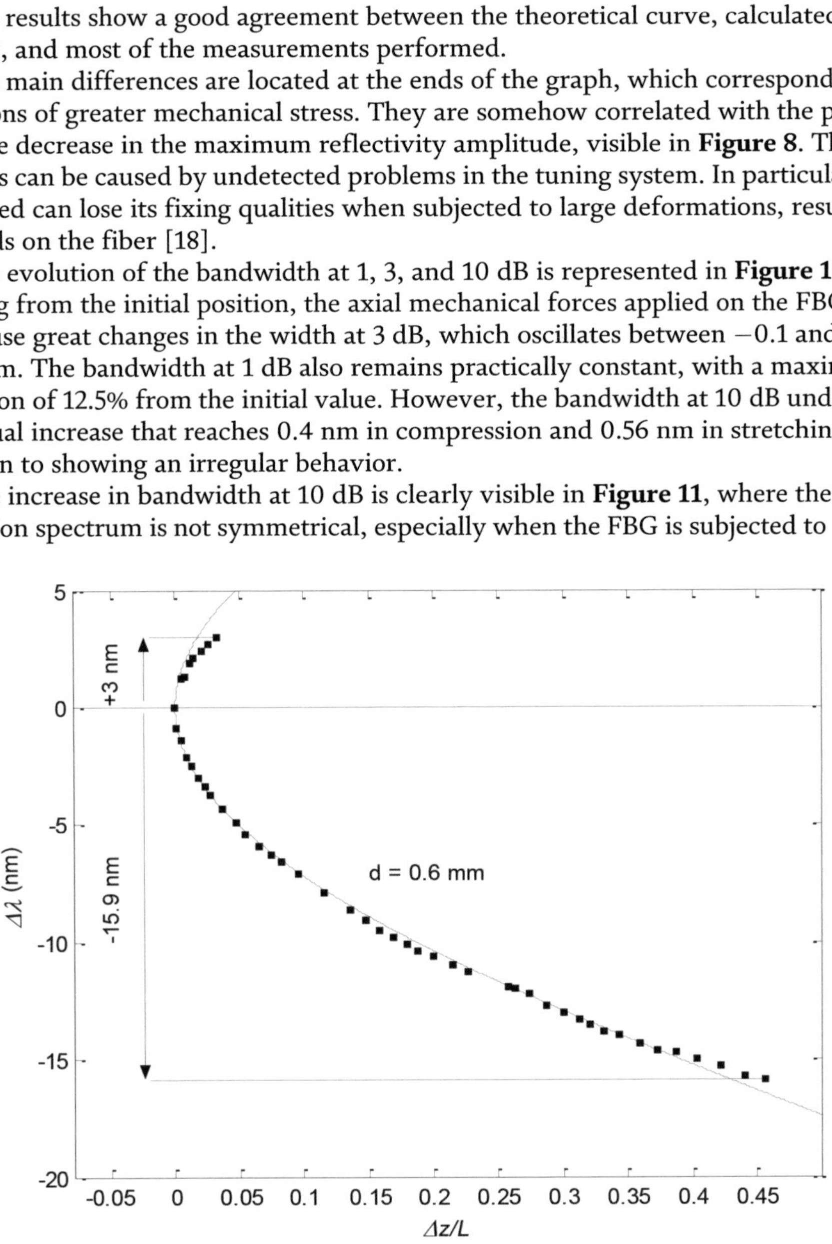

Figure 9.
Wavelength variation as a function of normalized horizontal displacement. (Dashed line: theoretical value, points: measured values).

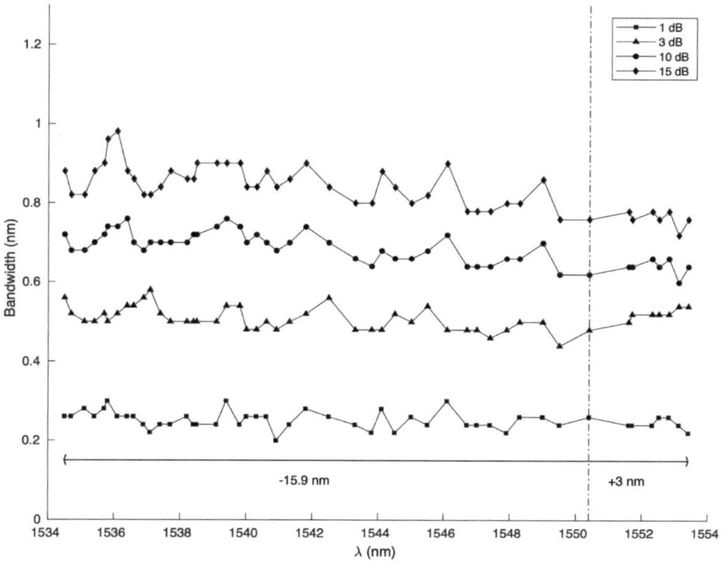

Figure 10.
Bandwidth evolution over the entire tuning range.

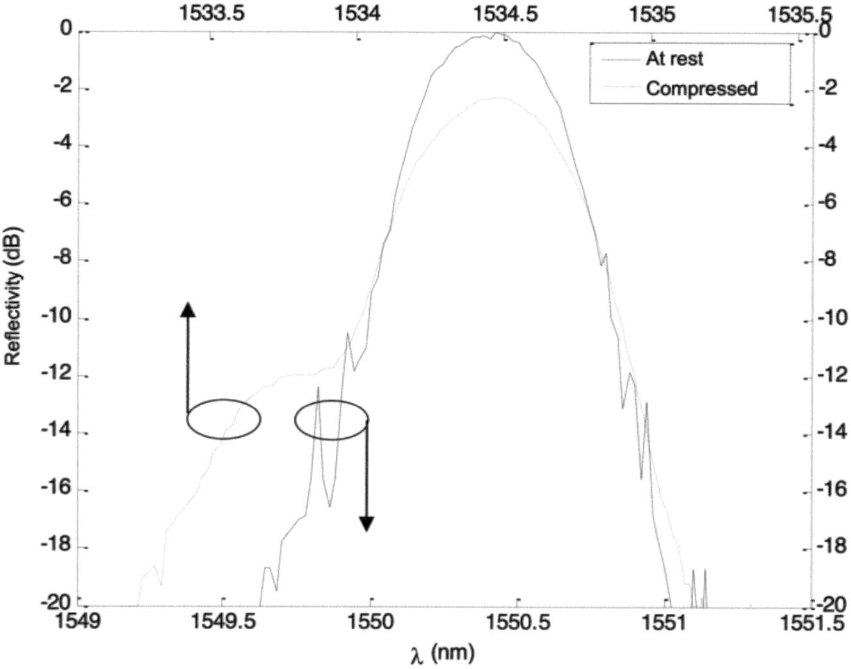

Figure 11.
Spectral response of a Bragg grating at rest (solid line) and under compression (dashed line).

compressions. Under these conditions, the pressure exerted along the grating is not uniform, which causes a variation in the grating period Λ and consequently a chirped spectrum. Consequently, the coupling between the propagation modes within the

FBG is no longer tuned to just one wavelength, thus decreasing the maximum reflectivity amplitude, and increasing the width at the base of the reflection spectrum.

On the other hand, the irregularities observed are largely due to the noise level being very close to 10 dB, causing sudden and profound variations in the reflectivity value.

According to **Eq. (7)**, the tension exerted on a Bragg grating, fixed at a distance d from the flexible blade, makes it possible to increase the tuning range without changing the length L. Theoretically, it is possible to move the reflection spectrum by 50 nm, even for small Δz deviations [7].

Another tuning test was performed with a Bragg grating with a reflectivity peak at 1546.3 nm, 3 dB spectral width equal to 0.46 nm and length 1.5 cm. The FBG was inserted into silicone, 6 mm away from the surface of a flexible acrylic base 16 cm long.

Figure 12 shows the relationship between the tuning range and the normalized travel distance. The measured tuning points do not follow the theoretical curve for d = 6 mm. The main cause for this deviation is a cause of the elastic properties of silicone. The curvature of the acrylic base is not followed by the silicone surface where the Bragg grating is inserted. Thus, the resulting tuning is much smaller than expected.

This problem is accompanied by a progressive decrease in the maximum reflectivity amplitude and in the broadening of the reflection spectrum, visible in **Figure 13**. The FBG reflectivity decreases by 3 dB over approximately 8 nm of tuning. The spectral widths at 1, 3, and 10 dB increase by 0.07, 0.29, and 1.33 nm, respectively.

4.2 Temperature tuning

To verify the dependence of the central reflection wavelength of the Bragg grating on temperature, the reflectivity spectrum of a grating centered initially at 1548.11 nm,

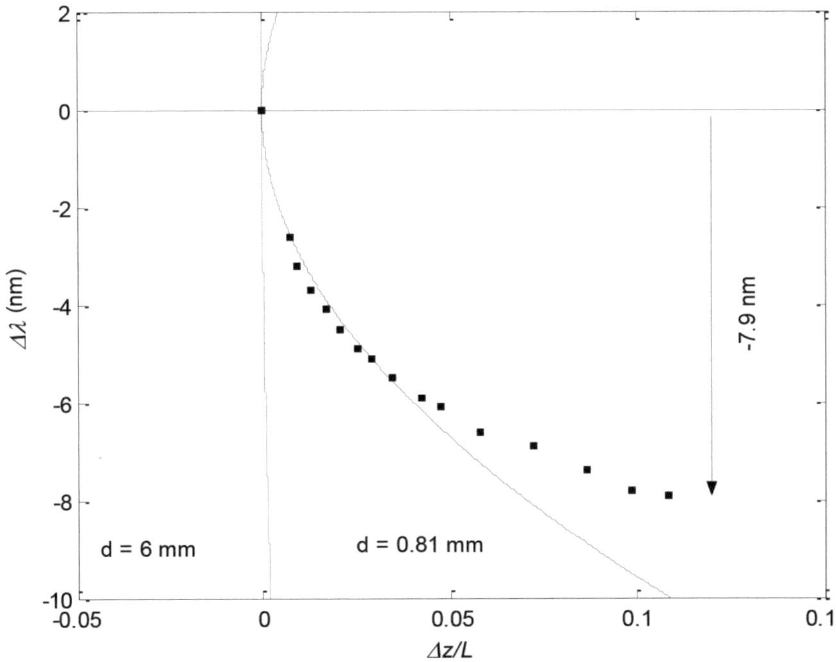

Figure 12.
Wavelength variation as a function of normalized horizontal displacement. (Dashed line: theoretical values, points: measured values).

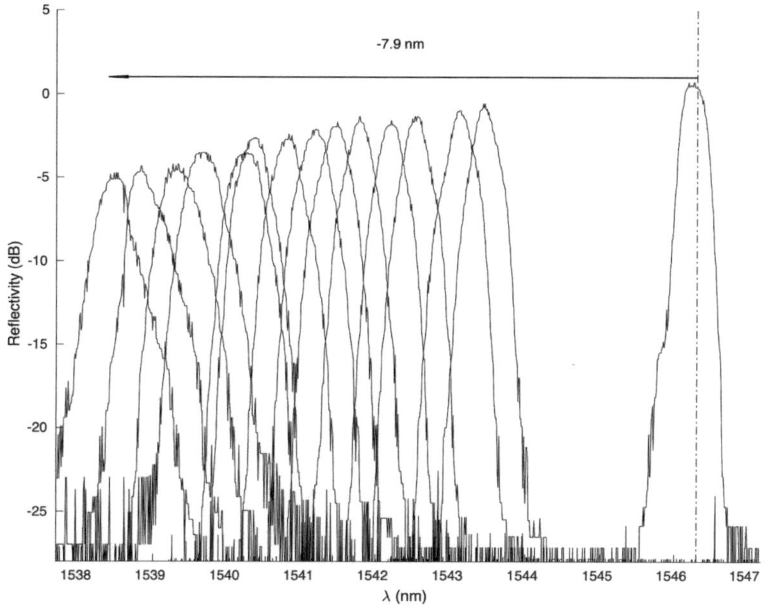

Figure 13.
Bragg grating tuning at 7.9 nm.

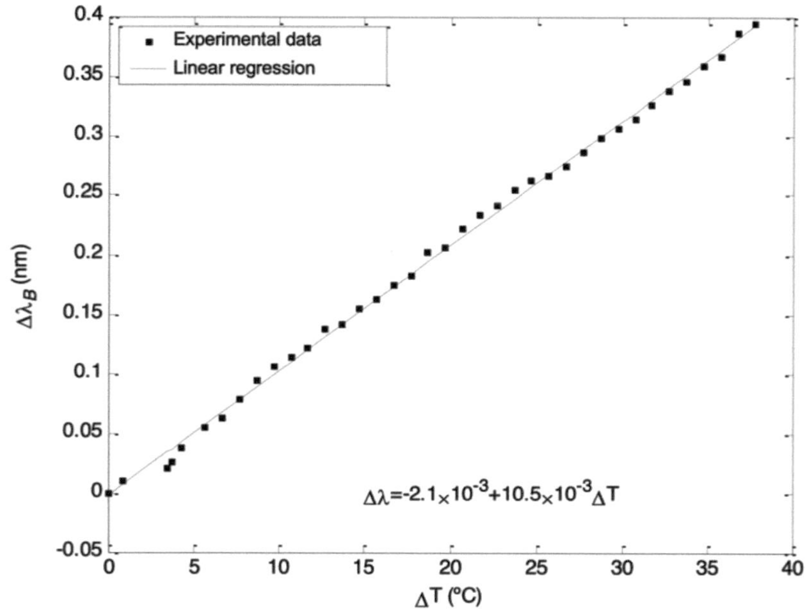

Figure 14.
Bragg wavelength variation as a function of FBG temperature (dashed line: theoretical value, points: measured values).

at a temperature of 14.3°C, was measured. The tuning method is based on the procedure described in Section 2.3. **Figure 14** shows the tuning results achieved at the expense of temperature variation in the Bragg grating. The tuning range accomplished was 0.4 nm, for a temperature variation of 40°C, and follows a linear relation between temperature and central wavelength variations.

5. Conclusion

In this chapter, two tuning processes for fiber Bragg gratings were presented: By mechanical stress on the fiber grating, or by changing its temperature.

In the first method, the optical fiber is more resistant to compression than to extension. The tuning range of the Bragg grating reflects this behavior, and therefore the Bragg grating must be produced with an initial central wavelength close to the upper tuning limit to increase the effective tuning wavelength span.

In the bending compression method, the material supporting the Bragg grating must be elastic enough to allow it to bend but must have the necessary rigidity to follow the curvature of the acrylic base. Therefore, silicone is not a suitable option for this tuning process, and other material with appropriate properties should be investigated.

The temperature tuning process achieved a poor tuning range (0.4 nm) compared to the mechanical tuning processes (19 nm), and it requires a permanent current source so that the Peltier cells can maintain the fiber grating temperature constant. The mechanical tuning system does not require the use of bulky and expensive current sources and maintains the desired tuning even if the electrical system is turned off. This is an additional advantage, which partially reduces electrical energy consumption. Furthermore, the spectral bandwidth (1, 3, and 10 dB) is maintained over a broad tuning range, but extreme bending or stretching of the fiber may chirp the Bragg grating and increase its spectral bandwidth.

Author details

Rogério Dionísio
Polytechnic Institute of Castelo Branco, DiSAC – Digital Services, Applications and Content, Castelo Branco, Portugal

*Address all correspondence to: rdionisio@ipcb.pt

References

[1] Hill KO, Fujii Y, Johnson DC, Kawasaki BS. Photosensitivity in optical fiber waveguides: Application to reflection filter fabrication. Applied Physics Letters. 1978;**32**(10):647-649

[2] Meltz G, Morey WW, Glenn WH. Formation of Bragg gratings in optical fibers by a transverse holographic method. Optics Letters. 1989;**14**(15):823-825

[3] Hill KO, Meltz G. Fiber Bragg grating technology fundamentals and overview. Journal of Lightwave Technology. 1997; **15**(8):1263-1276

[4] Russell PS, Archambault JL, Reekie L. Fibre gratings. Physics World. 1993;**6** (10):41-46

[5] Li C, Tang J, Cheng C, Cai L, Yang M. FBG arrays for quasi-distributed sensing: A review. Photonic Sensors. 2021;**11**(1): 91-108

[6] Ahlawat D, Arora P, Kumar S. Performance evaluation of proposed WDM optical link using EDFA and FBG combination. Journal of Optical Communications. 2019;**40**(2):101-107

[7] Giles CR. Lightwave applications of fiber Bragg gratings. Journal of Lightwave Technology. 1997;**15**(8):1391-1404

[8] Radzi NM, Latif AA, Ismail MF, Liew JY, Awang NA, Lee HK, et al. Tunable spacing dual-wavelength Q-switched fiber laser based on tunable FBG device. Photonics. 2021;**8**(12):524 MDPI

[9] Singh H, Sheetal A. Suitability of FBG for gain flatness of 64 $\times$ 10 Gbps DWDM system using hybrid (EDFA + YDFA) optical amplifier in C + L band up to 50 GHz (0.4 nm) channel spacing. Journal of Optical Communications. 2019. DOI: 10.1515/joc-2019-0134

[10] Othonos A, Kalli K, Kohnke GE. Fiber Bragg gratings: Fundamentals and applications in telecommunications and sensing. Physics Today. 2000;**53**(5):61

[11] Hunter DB, Minasian RA. Programmable high-speed optical code recognition using fibre Bragg grating arrays. Electronics Letters. 1999;**35**(5): 412

[12] Cheng HC, Wijanto E, Lien TC, Lai PH, Tseng SP. Multiple access techniques for bipolar optical code division in wireless optical communications. IEEE Access. 2020;**8**: 83511-83523

[13] Boukricha S, Ghoumid K, Mekaoui S, Ar-Reyouchi E, Bourouina H, Yahiaoui R. SAC-OCDMA system performance using narrowband Bragg filter encoders and decoders. SN Applied Sciences. 2020;**2**(6):1-9

[14] McGeehan JE, Hauer MC, Willner AE. Optical header recognition using fiber Bragg grating correlators. IEEE LEOS Newsletter. 2002;**16**(4):29-32

[15] Agrawal GP. Applications of Nonlinear Fiber Optics. London Elsevier: Ap, Academic Press; 2021

[16] Othonos A, Kalli K. Fiber Bragg gratings: Fundamentals and Applications in Telecommunications and Sensing. Boston, Mass.: Artech House; 1999

[17] Ibsen M, Set SY, Goh GS, Kikuchi K. Broadband continuously tunable all-fiber DFB lasers. IEEE Photonics Technology Letters. 2002;**14**(1):21-23

[18] Iocco A, Limberger HG, Salathe RP, Everall LA, Chisholm KE, Williams JAR, et al. Bragg grating fast tunable filter for wavelength division multiplexing. Journal of Lightwave Technology. 1999; **17**(7):1217-1221

[19] Goh CS, Mokhtar MR, Butler SA, Set SY, Kikuchi K, Ibsen M. Wavelength tuning of fiber Bragg gratings over 90 nm using a simple tuning package. IEEE Photonics Technology Letters. 2003;**15**(4):557-559

[20] Mokhtar MR, Goh CS, Butler SA, Set SY, Kikuchi K, Richardson DJ, et al. Fibre Bragg grating compression-tuned over 110 nm. Electronics Letters. 2003; **39**(6):509

[21] Set SY, Dabarsyah B, Goh CS, Katoh K, Takushima Y, Kikuchi K, et al. A widely tunable fiber Bragg grating with a wavelength tunability over 40 nm. InOptical Fiber Communication Conference 2001 Mar 17 (p. MC4). Optical Society of America

[22] Mokhtar MR, Ibsen M, Teh PC, Richardson DJ. Simple dynamically reconfigurable OCDMA encoder/ decoder based on a uniform fiber Bragg grating. InOptical Fiber Communication Conference 2002 Mar 17 (p. ThGG54). Optica Publishing Group

[23] Ohn MM, Alavie AT, Maaskant R, Xu MG, Bilodeau F, Hill KO. Dispersion variable fibre Bragg grating using a piezoelectric stack. Electronics Letters. 1996;**32**(21):2000

[24] Dabarsyah B, Goh CS, Khijwania SK, Set SY, Katoh K, Kikuchi K. Adjustable dispersion-compensation devices with wavelength tunability based on enhanced thermal chirping of fiber Bragg gratings. IEEE Photonics Technology Letters. 2003;**15**(3): 416-418

[25] Dionísio RP. Módulo de Posicionamento Micrométrico: Projecto, Construção e Caracterização. 1° Workshop de Investigação e Desenvolvimento em Engenharia Eletrotécnica e das Telecomunicações [Internet]. 2004 Feb 18 [cited 2022 Jun 27]. pp. 43-48. Available from: http:// hdl.handle.net/10400.11/8003

[26] de Brito André PS. Componentes optoelectrónicos para redes fotónicas de alto débito (Doctoral dissertation, Universidade de Aveiro (Portugal))

[27] Nogueira RN. Redes de Bragg em fibra óptica (Doctoral dissertation, Universidade de Aveiro (Portugal))

Chapter 5

A Review of Optical Interferometry Techniques for Quantitative Determination of Optically Active Materials in a Solution

Rahim Ullah, Raja Yasir Mehmood Khan and Muhammad Faisal

Abstract

Human diet is primarily comprised of optically active ingredients like glucose, sucrose, fructose, amino acids, lactic acid, cholesterol etc. Quality control is one of the most important processes in food industries to test, measure and verify the product for quality control standards. Optical techniques are mostly adopted in these industries for standardization of purity and concentration of optically active ingredients in their products. Quantitative measurements of optically active materials (OAMs) in a solution by interferometry have attracted the intention in present days due to their wide working range, high sensitivity and lower limit of detection. OAMs cause rotation of the angle of polarization when a plane-polarized light passes through them. The angle of rotation is distinct for different materials at different concentrations. For interferometric quantitative determination, the OAMs are generically placed in an arm of the interferometer and their effect on the interference fringe patterns are monitored as a function of their concentrations. Furthermore, the refractive indices of OAMs varies with their concentrations which directly affect the resultant interference pattern. Owing to the vast range of interferometric arrangements and processing techniques, this review assesses the different approaches adopted in detection of concentration of OAMs in a solution by interferometry.

Keywords: optically active materials, interferometry, refractive index, limit of detection, sensitivity

1. Introduction

Optically active materials are the fundamental constituents of the human body and its nutrients. Major part of the daily human diet is comprised of OAMs, and has a strong impact upon wellness and performance of the human body. The monitoring of OAMs in everyday nutrition is of vital importance for being healthy, lean and active. Therefore, a strong check of quality control is always the matter of concern is the industries including food, chemical, pharmaceutical, beverages etc. The OAMs cause the rotation in the angle of polarization of linearly polarized light when it pass through them. The variation in plane of polarization is different for different materials at different concentrations. OAMs are not limited to sugar (glucose, sucrose, fructose, maltose etc.), proteins, acids (tartaric acid, lactic acid etc.), cholesterol, etc.

A number of prescribed analytical techniques have been employed such as Ultra-voilet visible (UV–vis) absorption spectroscopy [1, 2], thin-layer chromatography (TLC) [3], infrared (IR) and Fourier transform infrared (FTIR) spectroscopy [4, 5] and Raman spectroscopy [6, 7]. However, optical techniques such as polarimetry [8, 9], interferometry [10, 11] and refractrometry [12] are commonly used in most of the practical applications for quantitative determination of OAMs due to their rapidity, noninvasiveness and non-destructive nature of their method of analysis.

OAMs are usually comprised of at least one asymmetric atom inside their molecular structure. The list of those atoms include carbon, sulfur, phosphorous, silicon etc. The asymmetric nature of these molecules result in the formation of two different types of isomers. The isomer of the same substance which rotates the plane of polarization of the light clockwise is called dextrorotatory or right-handed. However, those molecules which cause anti clock wise rotation of the polarization is called levorotatory or left-handed. Optical activity is the result of left–right asymmetry around the central carbon atom in the case of amorphous substances. The geometrical shape and chemical composition of both the molecules are same but left-handed isomer is mirror image of right-handed isomer and both of them are called enantiomers, as shown **Figure 1** for the D and L configuration of glucose molecule. Both of the enantiomorphs rotate the plane of polarization of light exactly by same magnitude but in opposite directions.

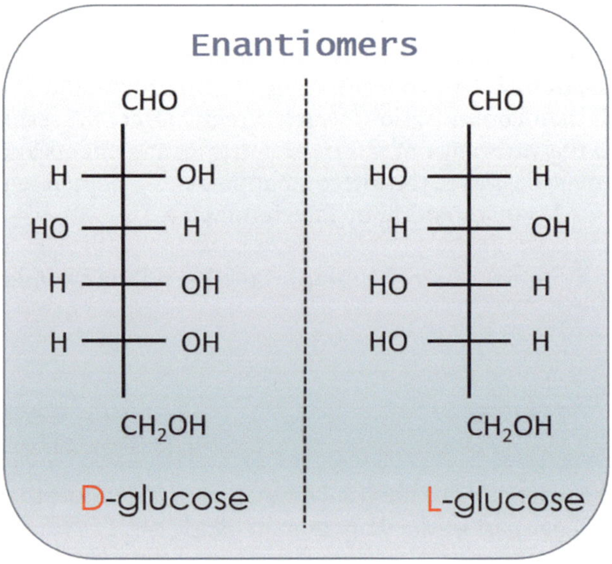

Figure 1.
Dextrorotatory and levorotatory configuration of glucose molecule.

2. Theoretical analysis

2.1 Biot's law

The foundation of detection of analytes concentration by using optical polarimetry trace back to the observation of Biot's in early nineteenth century [13]. The mathematical equation for the optical interaction of linearly polarized light with optically active specimen was described as below which is also called Biot's law;

$$[\alpha]_{\lambda}^{T} = \frac{\alpha}{LC} \tag{1}$$

The term $[\alpha]_{\lambda}^{T}$ is known as specific rotation of the OAM under consideration for specific wavelength of light (λ) at temperature (T). Where, α is the detected rotation of plane of polarization of light, L is the path length of the sample under test and C is the concentration of sample.

2.2 Analytical treatment of interference of linearly polarized light beams

Suppose two monochromatic light beams with same frequency originating from single source represented by the electric field vector $\vec{E}_1$ and $\vec{E}_2$ and interfere at the point of observation P. Let the light beams are generalized by the equations below however, physical waves will be represented by their real parts only

$$\vec{E}_1 = \vec{E}_{01} e^{-i(\omega t - k_1.r - \phi_1)} \tag{2}$$

$$\vec{E}_2 = \vec{E}_{02} e^{-i(\omega t - k_2.r - \phi_2)} \tag{3}$$

Where ω is the angular frequency of the monochromatic light wave, r is the position vector of incident point P, k_1 and k_2 are the wave vectors and ϕ_1 and ϕ_2 are the phase differences of beam 1 and 2, respectively. According to the principle of superposition, the irradiance at point 'P' can be calculated as:

$$I_P = \vec{E}_1 + \vec{E}_2 \tag{4}$$

The time averaged irradiance is proportional to the square of the amplitude of the electric field, i.e., for linear, homogeneous, isotropic dielectric medium;

$$I = \epsilon \upsilon \langle E^2 \rangle_T \tag{5}$$

Where ϵ is the permittivity of the medium and υ is the velocity of light in that medium. Considering the relative irradiance within the same medium except for the constant of proportionality, Eq. (4) may be transformed into;

$$I_P = \vec{E}_P.\vec{E}_P^{\,*} = \left(\vec{E}_1 + \vec{E}_2\right).\left(\vec{E}_1 + \vec{E}_2\right)^* = I_1 + I_2 + I_{12} \tag{6}$$

Where,

$$I_1 = \vec{E}_1.\vec{E}_1^{\,*} = E_{01}^2 \tag{7}$$

$$I_2 = \vec{E}_2 . \vec{E}_2^{\,*} = E_{02}^2 \tag{8}$$

and

$$I_{12} = \vec{E}_1 . \vec{E}_2^{\,*} + \vec{E}_2 . \vec{E}_1^{\,*} = \vec{E}_{01} . \vec{E}_{02} \left(e^{i\delta} + e^{-i\delta} \right) = 2\vec{E}_{01} . \vec{E}_{02} \cos \delta \tag{9}$$

Where,

$$\delta = \left(\vec{k}_1 - \vec{k}_2 \right) . \vec{r} + (\phi_1 - \phi_2) \tag{10}$$

The I_1 and I_2 are the irradiance of the individual beams and I_{12} represents the interference term. The symbol, δ represents the total phase difference between the two waves at the point of interest P. From Eq. (9) it can be observed that maximum inference will occur if both the beams are parallel to each other and no interference will occur if both the vectors are orthogonal to each other. For maximum interference Eq. (9) can be written as;

$$I_{12} = 2E_{01}E_{02} \cos \delta = 2\sqrt{I_1 I_2} \cos \delta \tag{11}$$

and total irradiance at point P will become

$$I_P = I_1 + I_2 + 2\sqrt{I_1 I_2} \cos \delta \tag{12}$$

The interference is called fully constructive if $\cos \delta = 2n\pi \textit{ where } n = 0, 1, 2, \ldots$ then Eq. (12) can be written as;

$$I_P = I_1 + I_2 + 2\sqrt{I_1 I_2} \tag{13}$$

Eq. (13) is called the maximum intensity or the maxima of fringes pattern (constructive interference) and can be interpreted as;

$$I_{max} = (E_{01} + E_{02})^2 \tag{14}$$

and I_P will be minimum if the term $2\sqrt{I_1 I_2} \cos \delta$ is negative, i.e., $\cos \delta = -1$ which is possible only if $\cos \delta = (2n + 1)\pi$ for n = 0,1,2, ... , and I_P will be called I_{min};

$$I_{min} = I_1 + I_2 - 2\sqrt{I_1 I_2} = (E_{01} - E_{02})^2 \tag{15}$$

If the two interfering light waves are mutually coherent then the time averaged value of $\cos \delta$ overtime period T should not vanish and a stationary fringe pattern will be obtained in space i.e.,

$$\langle \cos \delta \rangle_T = \frac{1}{T} \int_0^T \cos \delta \, dt \neq 0 \tag{16}$$

At different points of observation ($\vec{r}$), different values of $\langle \cos \delta \rangle_T$ will be obtained and resultantly different intensities will be obtained at different locations in space. Also $(\phi_1 - \phi_2)$ should not vary in time otherwise $\langle \cos \delta \rangle = 0$ and no sustained interference fringe pattern will be obtained. The quality of interference fringes is quantitatively described by visibility (V):

$$V = \frac{I_{max} - I_{min}}{I_{max} + I_{min}} \tag{17}$$

Where I_{max} and I_{min} represents the irradiances corresponding to maximum and adjacent minimum in the interference fringes. If one beam is incident with some small angle θ with respect to the plane of incidence then visibility could be defined as;

$$V = \frac{2\sqrt{I_1 I_2}\cos\psi}{I_1 + I_2} \tag{18}$$

Where, I_1 and I_2 are the irradiances of reference and sample beams respectively and ψ is the polarization angle of the sample beam of the interferometer [14].

3. Interferometry for OAMs detection

3.1 Mach-Zehnder interferometer

The Mach-Zehnder interferometer is most commonly used for sensing applications and was first introduced by Ludwig Zehnder in 1891 and Ludwig Mach in 1892 independently. In this interferometer, a coherent light beam is split into two using beam splitter and then recombined on another beam splitter with the help of two mirrors to obtain interference pattern. One part of the splitted beam is called reference while other one is called sensing arm. The chiral sample is kept in the sensing arm of the interferometer and its effect on the contrast of the interference fringes is detected. The interference fringes are normally recorded by a camera and analyzed by image processing techniques. MZI has good potential for the detection of OAMs therefore, most of the time exercised in the literature for this purpose.

Calixto et al. proposed a Mach-Zehnder interferometer (MZI) based wavefront division polarimeter for the measurements of chiral solute concentrations in solutions as shown in **Figure 2**. An optically active solution was kept in a sample chamber in the sensing arm of a MZI. One beam of the polarized light was passed through a liquid sample containing dissolved OAM while the other half of the light was used as reference beam. The plane of polarization of the incident linearly polarized light was rotated when the sample beam propagated through the chiral solutions. As a result, a decrease in the visibility of the interference pattern was observed with increase in the concentration of the OAMs in the solution. Contrast of the interference fringe pattern was maximum when both the sample and reference beams presented a polarization perpendicular to the plane of incidence. However, the visibility of the fringe pattern was deteriorated fully when the polarization of the sample beam is oriented parallel to the plane of incidence. The effect of decrease in the fringes visibility was mainly due to the increase in the refractive index of the solution with increase in the concentration of the OAMs as shown in **Figure 3**. The following calibration equation was obtained for variation of the visibility of the fringe pattern and concentration of the fructose solution:

$$V(c) = 0.643 - 0.31c^2 + 0.048\,c^4 \tag{19}$$

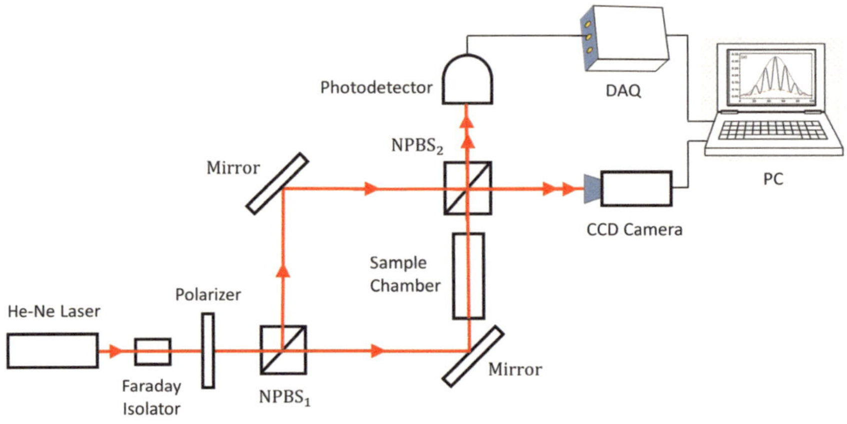

Figure 2.
Schematic diagram of the Mach-Zehnder interferometer and interference fringes pattern recording system [14].

Where, 'V' is the visibility of the interference fringe pattern and 'c' is the concentration of the chiral materials (**Figure 3**).

H.A. Razak et al. proposed an optical sensor based on fiber optic MZI for food composition detection as depicted in **Figure 4**. The MZI structure was employed as fiber optic sensor in single mode-multimode-single mode (SMS) structural configurations using fusion splicing technique. The interferometer was investigated with 4 cm and 8 cm sensing regions. The sensor response was tested for detection of water, sugar and oil from their respective refractive indices as representative major components of food. A red-shift was seen in the wavelength for increase in the refractive index of the constituent sample. The sensitivity of the sensor was found to be directly dependent on the length of the sensing region.

A miniature broad-band (BB) MZI was proposed by M. Kitsara et al. for the detection of label-free biochemical OAMs sensing as shown in **Figure 5**. A theoretical investigation was performed on Si-based MZI with BB input lights in the range 450 nm – 750 nm. They have proved that BB-MZI can be used as a miniaturized optical sensor with enhanced sensitivity, versatile biochemical sensing applications and economical fabrication and operating costs compared to its counterpart single wavelength (SW) MZI. Glucose was used as a representative biomolecular entity because of its relatively small size to demonstrate the designed BB-MZI to detect its concentration in a very diluted solutions with a higher efficiency. The phase changes of the evanescent field at Nd:YAG (532 nm) and He-Ne (633 nm) lasers were studied to evident that an optical setup could be designed where the source of light and MZI chip vary according to application. The theoretical transmission spectra of BB-MZI as a function of the refractive index of the solution were reported for 10 mM, 25 mM, 50 mM and 100 mM glucose concentrations. The recorded peak shifted with concentration of glucose. The performance of the BB-MZI was also observed with a hypothetical protein adlayer over the sensing arm of length 300 μm. An ultra-thin protein adlayer was sensed with recording of the spectral changes or by observing the variation in the integrated intensity. The potential of the designed interferometer was investigated for biomedical applications. It was observed that the performance of the BB-MZI is comparable to the SW-MZI without a requirement of a costly laser system as an input light source.

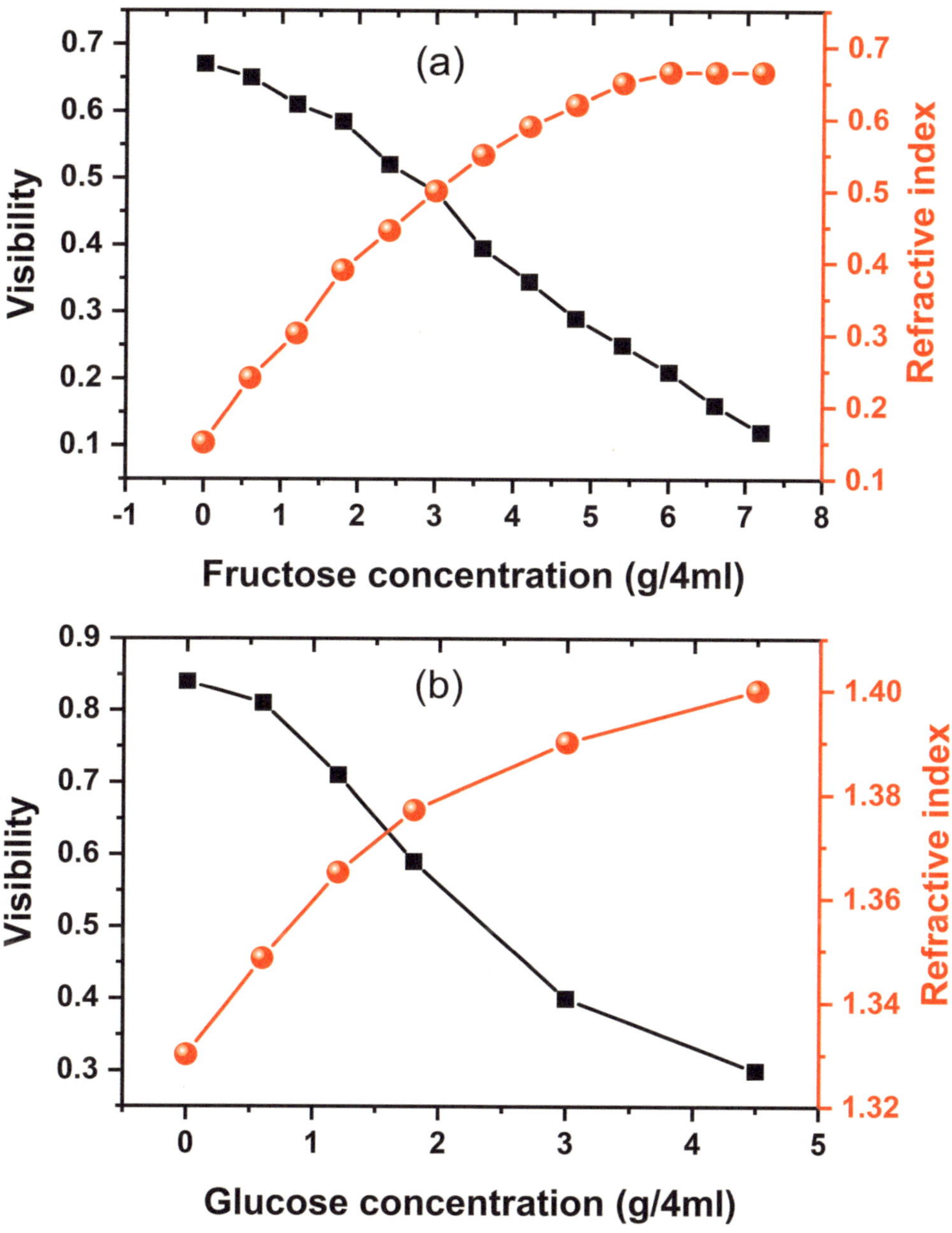

Figure 3.
The effect of concentration on the visibilities of fringes pattern by solutions of (a) fructose and (b) glucose [14].

A. Psarouli et al. also investigated a monolithically integrated BB-MZI for label free detection of biomolecules with high sensitivity as shown in **Figure 6**. A transducer based on monolithic silicon microphotonic was developed for this purpose. The MZI was fabricated from monomodal silicon nitride waveguides with silicone light emitting diodes (LEDs). BB light was injected into the interferometer setup and were sinusoidal modulated by optically active biomolecules with two different frequencies of the polarization before exiting the sensor. The distinct reporting of the two polarizations and simultaneous investigation of the TE and TM signals were performed by

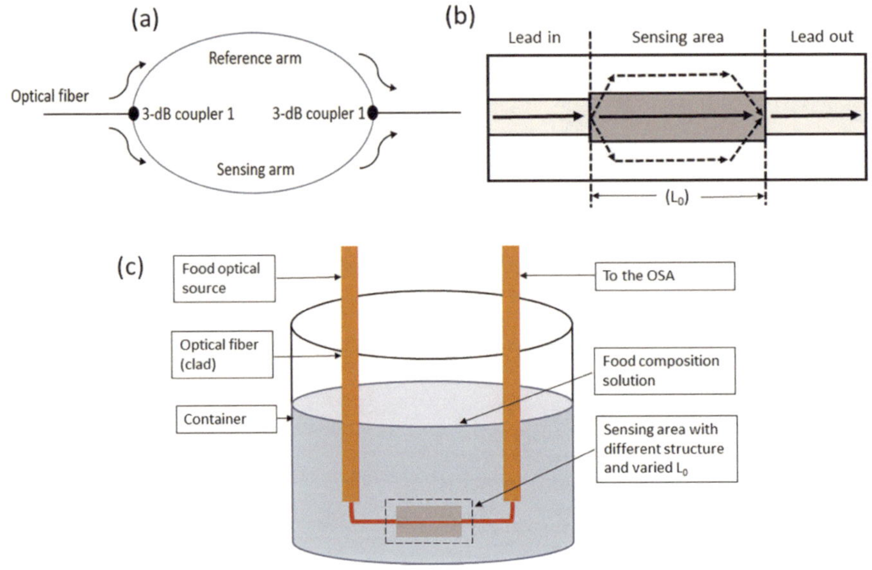

Figure 4.
(a) Basic block diagram of optical fiber MZ interferometer. (b) Schematic representation of SMS structure. (c) Schematic diagram of liquid concentration detector [15].

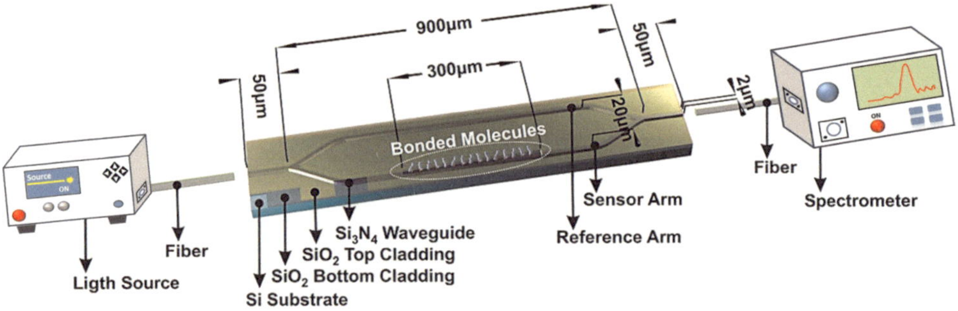

Figure 5.
Schematic diagram of the prosed BB-MZI for detection of label-free biochemical chiral materials sensing [16].

deconvolution in the Fourier transform. The quantitative determination of the binding adlyaers were made possible from their refractive indices by dual polarization analysis over the broad spectral range. The sensor was equipped with power and control electronics, a docking station, an off-chip fluidic circuit, a miniature spectrometer and an optical module. The set of ten interferometers were interrogated with a defined time delay by integrated LEDs which were operated by control electronics. The proposed interferometric sensor was found 60 and 550 times more sensitive than a two-lateral-mode spiral waveguide MZI [18] and polyimide-waveguide MZI [19], respectively.

An asymmetric Mach-Zehnder interferometer (aMZI) was introduced by M. J. Goodwin et al. for interferometric biosensing applications [11]. The device was manufactured using TriPleX technology. The interferometer was fabricated on a chip consist of Si_3N_4 waveguide with silica cladding which made a photonic integrated circuit (PIC). A sensing window was fabricated by locally removing the SiO_2 cladding which given a provision to analyte to make a contact with the waveguide. In the

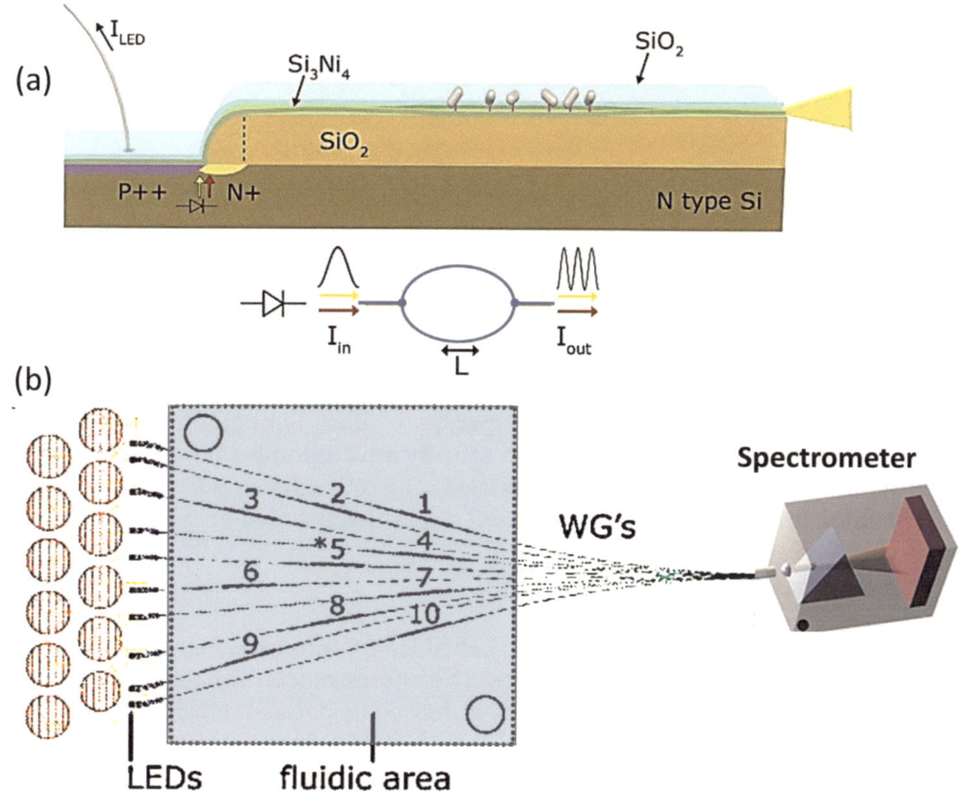

Figure 6.
(a) Schematic of the biochip showing the monolithic integration of the avalanche-type LED, the MZI, and the silicon nitride rib waveguide. (b) Layout of the 10 MZIs showing the MZI routes as well as the LED positions and metal [17].

proposed design, the incoming light was split into two arms by the waveguide. One of the arm was exposed to the analyte and the other arm was used as reference. The interference pattern was detected on at the point of recombination of the two arms due to the deliberately induced asymmetry. Subsequent to interaction of the analyte with the evanescent field of the waveguide at the sensing window, a phase shift was introduced due to variation in the refractive index. Performance of the proposed sensor platform in terms of signal-to-noise-ratio (SNR) and absolute response is compared with the commercial quartz crystal microbalance with dissipation (QCM-D). The aMZI proved itself dominated over the QCM-D due to measurement capability streptavidin binding with no need of the added complication of hydrodynamically coupled water which allow the elucidation of absolute protein adsorption. Also the aMZI presented 200 times good SNR and therefore offered a relatively lower limit of detection.

The operation of a versatile and sensitive integrated optical MZI biosensor with three-guide coupler at the output was demonstrated by B. J. Luff et al. as shown in **Figure 7**. The interferometric devices were designed by Ag^+–Na^+ ionexchange in glass substrates. The chemical modification of the waveguide surface of the interferometer made possible the detection of the biochemical species. The waveguide were designed in BGG36 glass with refractive index about 1.6 at 786 nm by Ag^+–Na^+ ion-exchange. Photolithographically patterned Ti film was used as the masking material with

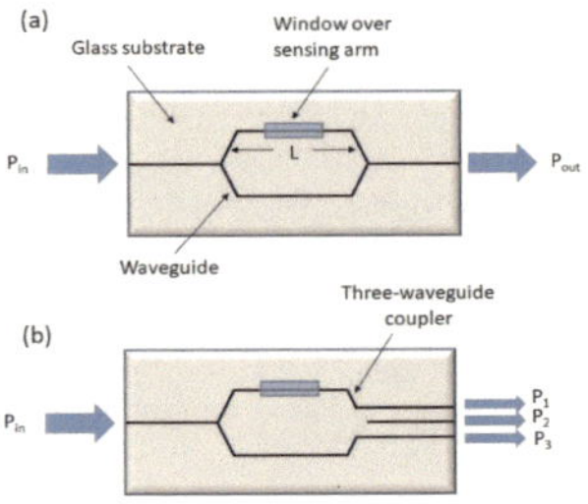

Figure 7.
(a) Schematic diagram of the biosensor device configuration based on MZI. (b) MZI with three-waveguide coupler [20].

opening width of 1 μm. The fabricated device was characterized by different concentrations of sucrose solutions to vary the superstrate index. For building up multilayers of protein over the sample surface, a high affinity interaction between vitamin biotin and protein streptavidin base system was used. The refractive index and thickness of the protein multilayer system was calculated reproducibly based on waveguide model.

3.2 Fabry–Pérot interferometer

Fabry–Pérot interferometer (FPI) also called etalon is based on an optical cavity made from two parallel reflecting surfaces. The interferometer is named after Charles Fabry and Alfred Perot for their invention in 1899. The parallel reflecting surfaces of the FP cavity is separated by a distance 'd' which allows the transmission of infinite number of parallel to each other as shown in **Figure 8**. A sharp constructive interference can be recorded when these parallel beam superimposed with each other. Free spectral range (FSR) analysis can be used to calculate the separation between the reflective ends of the cavity from refractive index information of a known medium. The spacing between the two reflective surfaces can be calculated from the FSR from refractive index of a known medium, as follow;

$$FSR = \Delta \upsilon = \upsilon_{m+1} - \upsilon_m = \frac{c}{2nd\cos\theta} \tag{20}$$

Where, υ is the frequency at which transmission of maximum intensity occurs, m determines an integral order of transmission peaks, θ is the angle of maximum transmission, d is the separation between the end reflective cavity surfaces and n is the index of refraction of a specific medium. FP interferometers are highly sensitive for the detection of OAMs therefore, mostly applied by the researchers in the literature.

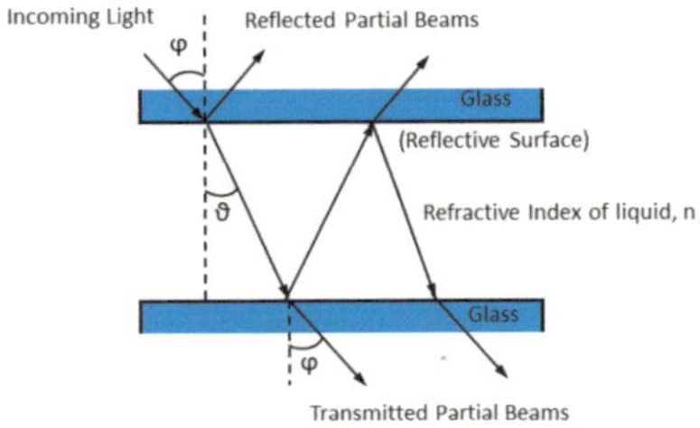

Figure 8.
Schematic diagram of the principle of the Fabry-Perot interferometer etalon. Basic structure of cavity [21].

The FPI is highly sensitive to any perturbation causing a variation in the optical path length between its two reflective mirrors as presented in **Figure 9**. Due to its compact size, high sensitivity and fast response, the FPI is applied for different physical parameters sensing, biosensing, gas sensing, current and magnetic field detection etc. [23]. FP etalons based optical sensors provide an efficient label-free biosensing capability with enhanced sensitivity. The biosensing of etalon is measured in terms of absorption or phase shift subsequent to interference between the reflected light beams from the two reflecting surfaces in its cavity [24–29].

G. Allison et al. investigated an efficient FP cavity coupled surface plasmon photodiode for electrical label-free biomolecular sensing [30]. The surface plasmonic sensor was developed inside a photovoltaic cell. The information of solutions containing biomolecules was extracted from its refractive indices in the form of electronic signal generated subsequent to incident light. The resultant photocurrent was enhanced due to surface plasmon mode coupling with the FP modes inside the photovoltaic cell due to its absorbing layer. An optically transparent substrate with special ability for surface plasmon resonance (SPR) was replaced by a silicone layer of semi-transparent optical nature. With the help of this mechanism, an absorbing layers was sandwiched between a metallic layer and an optically transparent conducting electrode. Photocurrent was caused as a result of incident light due to built-in

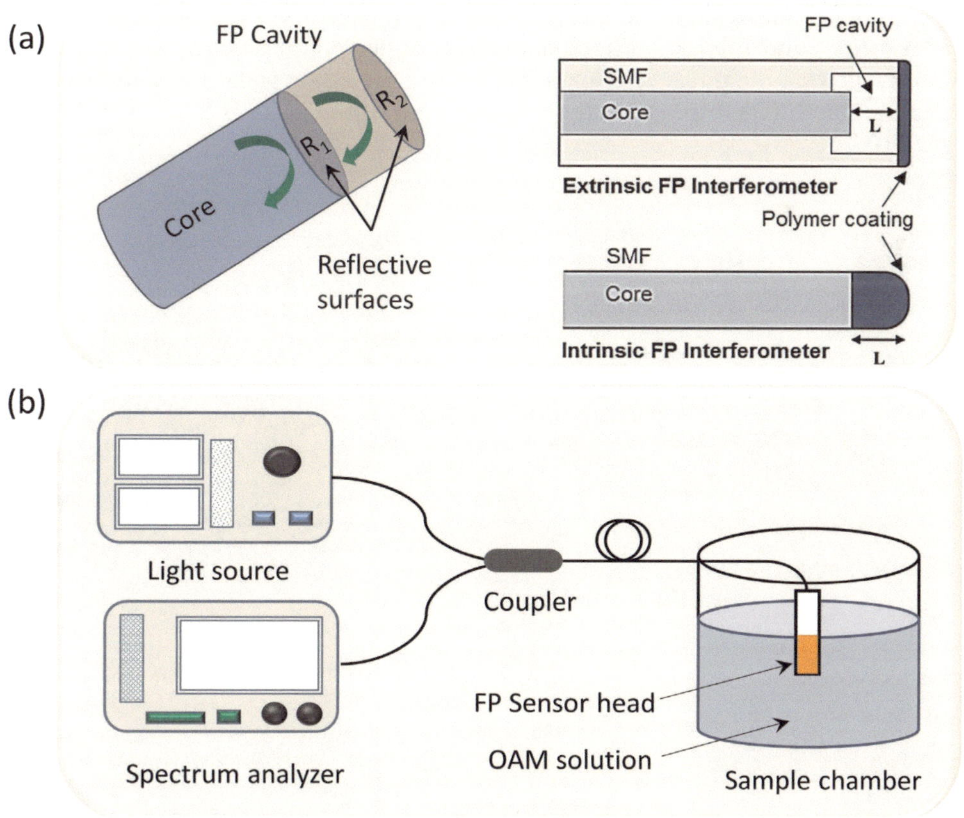

Figure 9.
(a) FP interferometer having reflective surfaces with reflectivity of R_1 and R_2, respectively. Examples of intrinsic and extrinsic FP interferometers. (b) Schematic representation of the FP based experimental setup employed for detection of gaseous biomolecules [22].

potential of fabricated device in a similar nature to that of photovoltaic cell. The surface plasmon was excited in the metal layer and generated the photocurrent simultaneously by illuminating a thin silicon layer by a visible light with single wavelength at a resonant angle. The photocurrent was reduced drastically by surface plasmon due to disruption of the distribution of electric field in the silicone layer. The mechanism were further enhanced by the silicone layer as an optical FP cavity to produce the FP modes which were coupled with the plasmon mode. The mechanism was confirmed by the simulation of the distribution of electric field which was further confirmed experimentally by electric detection of mode and resultantly the variations in the refractive index and the protein – protein interactions were measured.

A microfluidic optical sensor integrated with FP etalon geometry was investigated for detection of concentrations of different biochemical species in solution by K. E. You et al.. The concentration information were extracted in terms of the refractive index variation with concentration with high accuracy and sensitivity. The FP cavity was fabricated from a liquid channel with two partially reflected Ag/SiO_2 reflective surfaces. The refractive index dependent interference peaks were achieved in the transmission spectra subsequent to transmission of light through the fluid channel. Concentrations of different biomolecules, i.e., glucose, potassium chloride and sodium chloride were calculated from their refractive indices in terms of a shift in the position of maxima of wavelength of the interference peaks in the transmission spectra. The devised optical sensor shown a linear response with good accuracy. Sensitivities of 10^{-3}, $1.4x10^{-3}$ and $1.8x10^{-3}$ refractive indices per percent of glucose, KCl and NaCl, respectively were obtained. Schematic diagram of the investigated FP cavity based optical sensor and its response to the optically active glucose sample was shown in **Figure 10**.

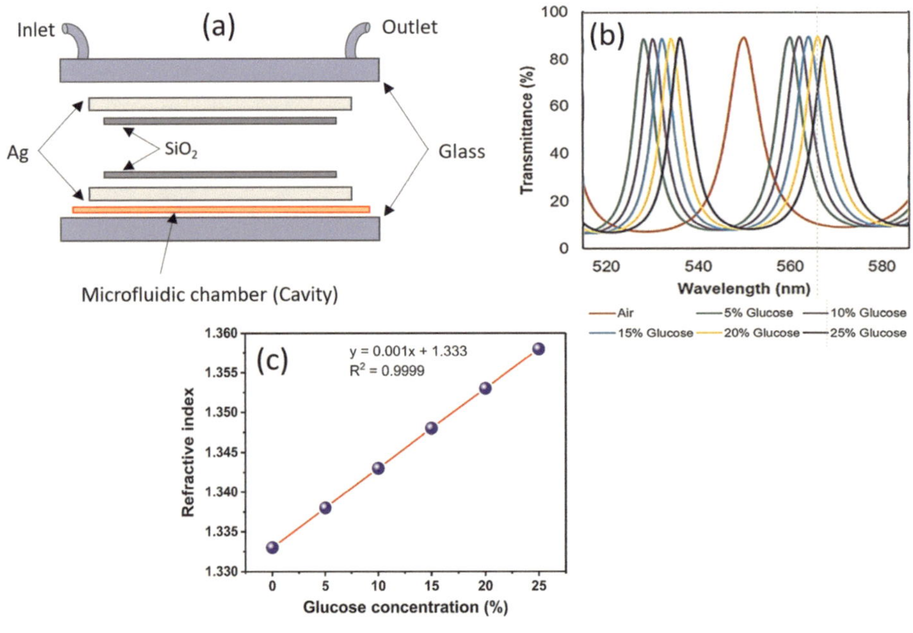

Figure 10.
Schematic diagram of the devised modified microfluidic Fabry-Perot etalon (a) 2D view, (b) simulation (c) refractive index vs. glucose concentration (%) [21].

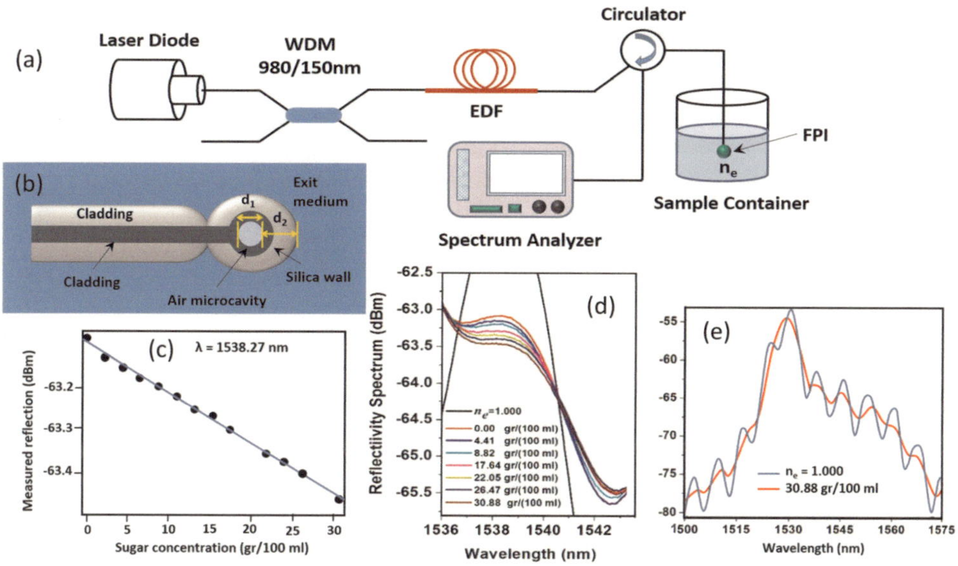

Figure 11.
(a) Schematic diagram of the experimental setup for MFPI based sugar concentration sensing, (b) schematic view of the fabricated MFPI, (c) reflected energy and the refractive index of the exit medium at λ = 1538.27 nm as a function of the sugar concentration (d) measured reflected power spectra for different sugar concentrations and (e) enlarged view of the spectra [31].

A micro Fabri-Pérot interferometer (MFPI) was designed and developed for quantitative determination of sugar in a transparent solutions by G. Chavez et al. as shown in **Figure 11(a)**. The MFPI was developed in the form of a micro bubble made of a hollow core photonic crystal fiber (PCF) as shown in **Figure 11(b)**. The cavity was fabricated by splicing of a segment of PCF to a single mode fiber (SMF) by a conventional arc fusion splicer which form an air MFPI. The fabricated MFPI then acted like an optical cavity having two reflecting surface of different refractive indices separated by a distance d. The erbium doped fiber (EDF) was illuminated by 200 mW laser diode with 980 nm wavelength with the help of wavelength division multiplexer (WDM). Output light from EDF was incident on the MFPI subsequent to passing through a three-port circulator. The reflection from MPI was guided towards optical spectrum analyzer (OSA) by port three of the circulator. The MFPI was immersed inside a sugar solution filled cuvette and the reflected interference patterns were recorded at different concentrations of sugar solution in the range 0–30.88 g/100 ml. The contrast of the interference fringes decreases with increased in the sugar concentration in solution as illustrated in **Figure 11(d)**. The predicted results from simulation were also experimentally confirmed with good agreement. The reflected optical power was directly decreased with increase in the concentration and resultant refractive index of solution with sensitivity of −0.0123 dBm/(g/100 ml) at 1538.27 nm wavelength.

J. Martini et al. also proposed a glucose concentration sensor in interstitial fluids based on a small size double-chamber FP etalon. One of the FP chamber of the proposed sensor was used as reference to overcome the effect of ambient temperature variations. The 400 μm etalon cavity was filled with water – glucose solution which had FSR of 680 pm in response to the normal incident light of wavelength 850 nm. A wavelength shift of ~1 pm was produces per 1 mg/dl of the optically active analyte (glucose). The light beam from an SM vertical cavity surface emitting laser (VCSEL,

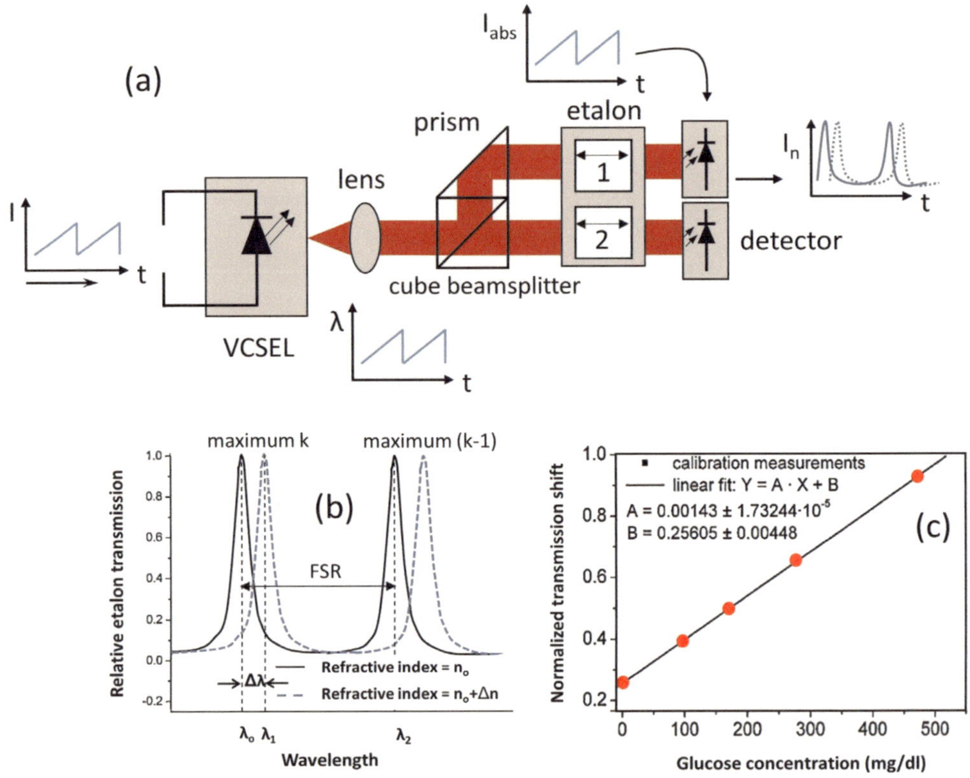

Figure 12.
(a) Schematic diagram of the experimental setup of the double-chamber FP etalon (b) spectral position of FP modes for two different refractive indices (c) calibration curve of the refractive index with varied concentration of glucose [32].

850 nm) was guided towards 50/50 beam splitter and was incident on one of the FP chamber subsequent to proper collimation. The perpendicular half of the beam was redirected into the other FP chamber with the help of a prism. The transmitted light signal from FP chamber were recorded by PIN photodiodes with two active segments as shown in **Figure 12**. Difference in the refractive indices of the two etalon chambers produced a phases in the transmitted optical signals. The proposed optical sensor was studied in the range 0–700 mg/dl of glucose concentration with precision of ±2.5 mg/dl. The temperature compensation was confirmed in the range 32–42°C.

3.3 Sagnac interferometer

In the Sagnac interferometer, the light beam is split to follow the same optical path but in opposite directions in the form of a closed loop. The beams get interference upon returning back to the point of entry. In the case of optical fiber Sagnac interferometer, a section of birefringent fiber is splices to the loop which causes interference between the counter propagating light beams.

A Sagnac interferometer based optical sensor system was designed, developed and demonstrated by T. Kumagai et al. for quantitative analysis of glucose in a solution. The optical rotation proportion to concentration of glucose concentration was measured in a Sagnac loop from a phase difference introduced between the clockwise

(CW) and counter clockwise circularly polarized light. The proposed optical sensor was composed of a Sagnac interferometer made from a polarization maintaining fiber (PMF) and other optical components to avoid the unwanted sources of noise due to different reasons. A coherent light of wavelength 840 nm from a super luminescent diode was transmitted in the sensor system. Two orthogonal mode of polarizations with a minor difference in their propagations constants were transmitted in the interferometer. The ambient sources of noise those were temperature variations and external vibrations may vary the zero level of the output signal which were controlled by fabricating the interferometer from PMF. A quarter-wave retarder was used to convert the linearly polarized light into circularly polarized light which was subsequently passed through a low birefringence span SMF or in free space. The interference signal achieved from recombination of CCW and CW lights was confirmed from the output of a preamplifier and was physically observed on an oscilloscope. The polarization measurement system (polarimeter) was checked by measuring the phase difference with the help of Faraday effect optical rotation measurement setup. In this regard, current was applied to a 1300-turn copper coil and a span fiber was a wounded around the coil. Concentration of glucose was measured from the degree of optical rotation using Biot's law as shown in Eq. (1). The phase difference analysis was performed in a I dm long measurement cell. The specific rotation of +51.6 and − 91.2 were measured for glucose and fructose respectively which were close to their physical properties. To make the sensor suitable for practical applications, the active length of the measurement cell was reduced to about 2 mm and the resolution of the sensor was monitored. A trial based noninvasive measurement with human body was performed by skin webbing between fingers. The resultant interference between CCW and CW lights was investigated. The skin webbing caused a bias in the interference signal which was observed in the form of noise due to a phase difference between CCW and CW light. A resolution of 1 mg/dl was achieved for glucose concentration and 0.5 mdeg resolution of optical rotation was detected for the devised sensor (**Figure 13**).

An optical polarimetry based Sagnac interferometry technique was investigated for noninvasive glucose sensing by A. M. Winkler et al. [34]. A phase sift in the interference fringes of the Sagnac interferometer was detected in a glucose solution as a representative OAM. The proposed method was linked with the sugar detection from the aqueous humor of human eye. The interferometer was simulated such that the counter propagating beams in which one of them passed through an optically active sample caused a difference in the optical path traversing. The effect was due to a difference in the refractive indices of the left and right circularly polarized beams.

A compact PCF Sagnac interferometer based glucose sensor was introduced by G. Ann et al. [35]. A light signal from a broadband light source was launched in a 3 dB coupler and split into two beams.. A Saganac loop was mainly comprised of a polarization controller and PCF spliced with an SMF. The splitted beams counter propagated and interfered with an accumulated phase difference when passed through the PCF. The interference signal was effected greatly by the phase difference between the two orthogonal guided modes of PCF. A similar trend of phase birefringence was observed when the wavelength was varied in the range 1000–2000 nm with a gradual decrease in the maxima of the curve with increase in the glucose concentration. The nature of the devised sensor was analyzed between 15 cm to 40 cm. It was observed that the response of the sensor was highly sensitive to the PCF length. A prominent interference peaks were observed between 1050 nm to 2000 nm. The interference

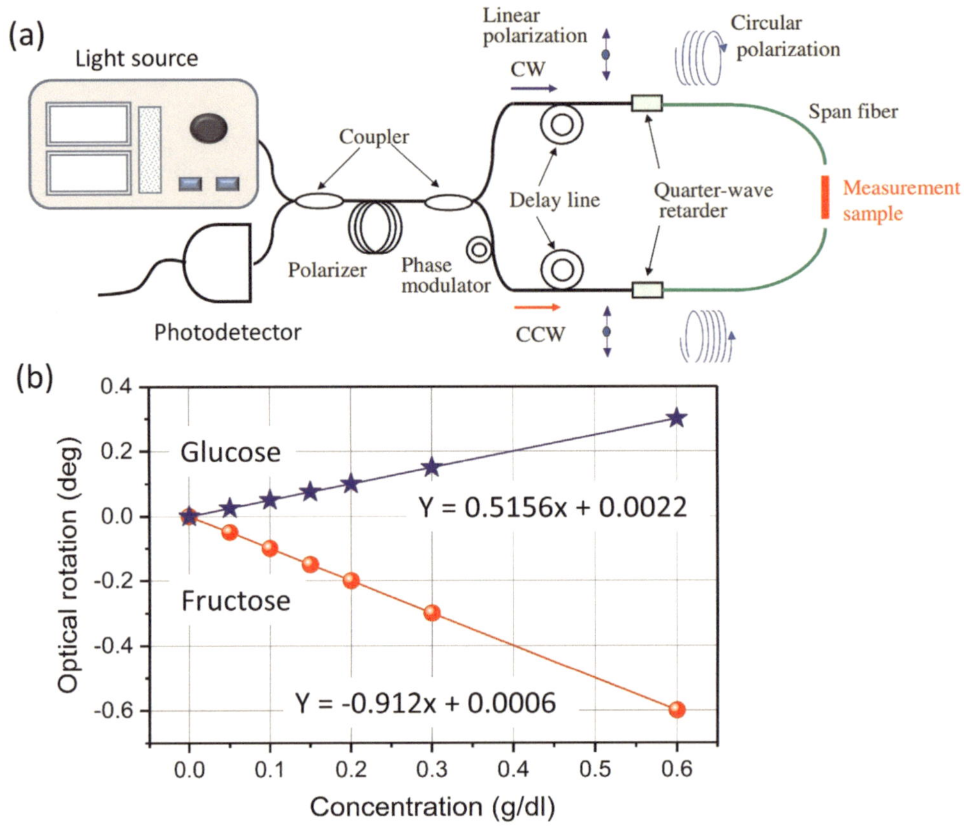

Figure 13.
(a) Configuration of optical rotation measurement system with a Sagnac interferometer. (b) Concentration dependent optical rotation of sugar samples [33].

signatures becomes highly sensitive with good SNR for 20 cm PCF length. An average sensitivity of 0.76 mg/dL of glucose solution in water was recorded which is lower than 70 mg/dl of hypoglycemia episodes. The sensor was designed for effective sensing of glucose level in the patients with hypoglycemia.

3.4 Michelson interferometer

In this type of interferometer, a coherent light beam is split into two by a beam splitter. Each of the two beams are reflected back and recombines at the same beam splitter to get interference pattern. Although, the Michelson interferometer has good potential for the detection of OAMs but rarely applied due to its relatively complex optical arranges. L. K. Chin reported a droplet Michelson interferometer for biochemical and protein detection. The interferometer was fabricated in the form of on-chip liquid grating as schematically shown in **Figure 14(a)**. A branch of the interferometer was spared for injection of two immiscible liquids. A T-junction was developed for the formation of the liquid grating. The other branch with microchannel was filled with immersion oil which caused a phase shift due to optical path difference produced in the paths of light transmitted through the core and the cladding. A buffer solution was injected in the third branch to measure its refractive index. An optical fiber was

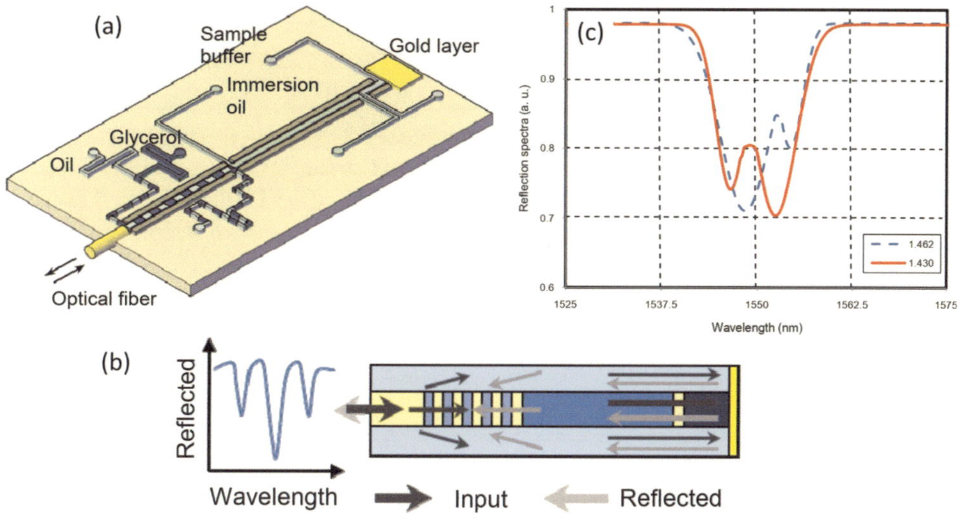

Figure 14.
Schematic illustrations of (a) the droplet Michelson interferometer and (b) the physics model of Michelson interferometer (c) reflection spectra of liquid grating Michelson interferometer [36].

aligned with one end of the microchannel for input/output light detection and the other end was coated with a gold film to use it as end mirror. The light was coupled from core to the cladding with the help of liquid grating. In the interferometer, the second and third branches of the microchannel was used to propagate the light which was reflected back by the gold layer which caused an optical path difference. An interference pattern was observed with attenuation band when both the light signals recombines in the core subsequent to passing through the liquid grating as shown in **Figure 14(b)**. The microchip was fabricated in polydimethylsiloxane (PDMS) using lithography. Sputtering technique was applied to coat the sidewall of the PDMS edge. The PDMS slab with the pattern was attached with the unpatented PDMS slab using plasma bonding. The broadband light was coupled by an optical circulator with the optofluidic chip. The reflection from the optofluidic interferometer was detected by an OSA. The immersion oil and glycerol with refractive indices 1.462 and 1.420, respectively were used as carrier and dispersed flow, respectively. Reflection from the devised interferometer for different buffers with distinct refractive indices was recorded by the spectrum analyzer as shown in **Figure 14(c)**. Two distinct attenuation peaks were observed in the overall attenuation band at slightly different positions and intensities.

4. Conclusion

In this study, different interferometric optical sensors employed for the detection and quantification of OAMs are briefly reviewed (**Table 1**). The working principle, design and the important performance parameters including working range, sensitivity, and limit of detection were discussed in detail. Different materials analyzed in the sensors were outlined with special focus on sugar and protein being a representative OAMs. Two different interferometric arrangements, i.e., free space and optical fiber based optical sensors were discussed. It was

Interferometer	Characterized material	Dynamic range	Sensitivity	Limit of detection	Reference
Mach-Zehnder	Protein	50–1000 nm	2000 nm/RIU	—	[11]
	Glucose	625 mg/ml	−2.36°/ml	25.5 mg/ml	[14]
	Fructose	922 mg/ml	−1.31°/ml	37.5 mg/ml	
	Sucrose	—	4.413 nm/RUI	—	[15]
	Glucose/protein	—	—	10 mM	[18]
	Protein	—	0.4 rads/ng/mm^2	2 pg/mm^2	[20]
	Glucose	0–25%	0.001 RUI/%	10-5 RUI	[21]
Fabry–Pérot	glycerol	0–50%	1.5 μA/mRIU	10 pg/ml	[30]
	Sugar	0–30.88 g/100 ml	−0.0123 dBm/(g/100 ml)	—	[31]
	Glucose	0–700 mg/dl	2.5 mg/dl		[32]
Sagnac	Glucose	0–0.6 g/dl	0.5156 deg/(g/dl)	1 mg/dl	[33]
	Fructose	0–0.6 g/dl	0.9120 deg/(g/dl)	—	
	Glucose	0–500 mg/dl	—	—	[34]
	Glucose	0–120 g/l	2.63 nm/g/l	0.76 mg/dl	[35]

Table 1.
Comparison of different OAMs analyzed by different interferometric arrangements.

concluded that optical fiber based interferometry has mind blowing potential for highly precise sensing of OAMs with good sensitivity and can be applied in industrial and research and development applications.

Author details

Rahim Ullah*, Raja Yasir Mehmood Khan and Muhammad Faisal
National Institute of Lasers and Optronics College, Pakistan Institute of Engineering and Applied Sciences, Islamabad, Pakistan

*Address all correspondence to: rahimullahphysics28@gmail.com

References

[1] de Souza RR, Fernandes DD d S, Diniz PHGD. Honey authentication in terms of its adulteration with sugar syrups using UV–vis spectroscopy and one-class classifiers. Food Chemistry. 2021;**365**:130467. DOI: 10.1016/J. FOODCHEM.2021.130467

[2] Chen C, Jiang Y, Kan J. A noninterference polypyrrole glucose biosensor. Biosensors & Bioelectronics. 2006;**22**(5):639-643. DOI: 10.1016/J. BIOS.2006.01.023

[3] Islam MK, Sostaric T, Lim LY, Hammer K, Locher C. A validated method for the quantitative determination of sugars in honey using high-performance thin-layer chromatography. JPC – Journal of Planar Chromatography – Modern TLC. 2020; **33**(5):489-499. DOI: 10.1007/ S00764-020-00054-9

[4] Khalil SKH, Kamil M. Application of FT-IR spectroscopy for rapid and simultaneous quality determination of some fruit products environmental microbiology view project drug delivery system view project. Natural Science. 2011;**9**(11):21-31

[5] And JFDK, Downey G. Detection of sugar adulterants in apple juice using Fourier transform infrared spectroscopy and Chemometrics. Journal of Agricultural and Food Chemistry. 2005; **53**(9):3281-3286. DOI: 10.1021/ JF048000W

[6] Guven B, Durakli-Vellioglu S, Boyaci İH. Rapid identification of some sweeteners and sugars by attenuated Total reflectance-Fourier transform infrared (ATR-FTIR), near-infrared (NIR) and Raman spectroscopy. Gıda. 2019;**44**(2):274-290. DOI: 10.15237/ GIDA.GD18119

[7] Xu Y et al. Raman spectroscopy coupled with chemometrics for food authentication: A review. TrAC Trends in Analytical Chemistry. 2020;**131**: 116017. DOI: 10.1016/J. TRAC.2020.116017

[8] Soni M. A review on the measurement of optical activity by using a Polarimeter. Journal of Pharmacognosy and Phytochemistry. 2019;**8**(2):358-360

[9] Pokrzywnicka M, Koncki R. Disaccharides determination: A review of analytical methods. Critical Reviews in Analytical Chemistry. 2017;**48**(3): 186-213. DOI: 10.1080/ 10408347.2017.1391683

[10] Tan C-T, Wang C, He J-N, Chiu M-H. Phase sensitive optical rotation measurement using the common-path heterodyne interferometry and a half-wave plate at a specific azimuth angle. OSA Continuum. 2021;**4**(1):239-251. DOI: 10.1364/OSAC.415766

[11] Goodwin MJ, Besselink GAJ, Falke F, Everhardt AS, Cornelissen JJLM, Huskens J. Highly sensitive protein detection by asymmetric Mach–Zehnder interferometry for biosensing applications. ACS Applied Bio Materials. 2020;**3**(7):4566-4572. DOI: 10.1021/ ACSABM.0C00491

[12] Buiarelli F, Di Filippo P, Pomata D, Riccardi C, Rago D. Determination of sugar content in commercial fruit juices by Refractometric, volumetric and chromatographic methods. Journal of Nutritional Therapeutics. 2016;**5**(3):75-84. DOI: 10.6000/1929-5634.2016.05.03.3

[13] Handapangoda, Chintha C, Premaratne M, Yeo L, Friend J. Laguerre Runge-Kutta-Fehlberg method for simulating laser pulse propagation in

biological tissue. IEEE Journal of Selected Topics in Quantum Electronics. 2008;**14**(1):105-112. DOI: 10.1109/JSTQE.2007.913971

[14] Calixto S, Martinez-Ponce G, Garnica G, Figueroa-Gerstenmaier S. A Wavefront division Polarimeter for the measurements of solute concentrations in solutions. Sensors. 2017;**17**(12):2844. DOI: 10.3390/S17122844

[15] Razak HA, Sulaiman NH, Haroon H, Mohd Zain AS. A fiber optic sensor based on Mach-Zehnder interferometer structure for food composition detection. Microwave and Optical Technology Letters. 2018;**60**(4):920-925. DOI: 10.1002/MOP.31072

[16] Kitsara M, Raptis I, Misiakos K, Makarona E. Broad-band mach-zehnder interferometry as a detection principle for label-free biochemical sensing. Proceedings of IEEE Sensors. 2008: 934-937. DOI: 10.1109/ICSENS.2008.4716594

[17] Psarouli A et al. Monolithically integrated broad-band Mach-Zehnder interferometers for highly sensitive label-free detection of biomolecules through dual polarization optics. Scientific Reports. 2015;**5**(1):1-11. DOI: 10.1038/srep17600

[18] Kee JS, Kim KW, Park MK, Liu Q, Gu Z. Single-channel Mach-Zehnder interferometric biochemical sensor based on two-lateral-mode spiral waveguide. Optics Express. 2014;**22**(23): 27910-27920. DOI: 10.1364/OE.22.027910

[19] Bruck R, Melnik E, Muellner P, Hainberger R, Lämmerhofer M. Integrated polymer-based Mach-Zehnder interferometer label-free streptavidin biosensor compatible with injection molding. Biosensors & Bioelectronics. 2011;**26**(9):3832-3837. DOI: 10.1016/J.BIOS.2011.02.042

[20] Luff BJ, Wilkinson JS, Piehler J, Hollenbach U, Ingenhoff J, Fabricius N. Integrated optical Mach-Zehnder biosensor. Journal of Lightwave Technology. 1998;**16**(4):583-592

[21] You KE, Uddin N, Kim TH, Fan QH, Yoon HJ. Highly sensitive detection of biological substances using microfluidic enhanced Fabry-Perot etalon-based optical biosensors. Sensors and Actuators B: Chemical. 2018;**277**:62-68. DOI: 10.1016/J.SNB.2018.08.146

[22] Khan S, Le Calvé S, Newport D. A review of optical interferometry techniques for VOC detection. Sensors and Actuators A: Physical. 2020; **302**:111782. DOI: 10.1016/J.SNA.2019.111782

[23] Huang YW, Tao J, Huang XG. Research Progress on F-P interference—Based Fiber-optic sensors. Sensors. 2016;**16**(9):1437. DOI: 10.3390/S16091424

[24] Que L, Giorno R, Zhang T, Gong Z. A nanostructured Fabry-Perot interferometer. Optics Express. 2010; **18**(19):20282-20288. DOI: 10.1364/OE.18.020282

[25] Dohi T, Matsumoto K, Shimoyama I. The optical blood test device with the micro Fabry-Perrot interferometer. 17th IEEE International Conference on Micro Electro Mechanical Systems. Maastricht MEMS 2004 Technical Digest. IEEE, 2004. pp. 403-406

[26] Bhatia V et al. Optical fibre based absolute extrinsic Fabry - Pérot interferometric sensing system. Measurement Science and Technology. 1996;**7**(1):58. DOI: 10.1088/0957-0233/7/1/008

[27] Majchrowicz D, Hirsch M, Wierzba P, Bechelany M, Viter R, Jędrzejewska-Szczerska M. Application of thin ZnO ALD layers in Fiber-optic Fabry-Pérot sensing interferometers. Sensors. 2016;**16**(3):416. DOI: 10.3390/S16030416

[28] Yeatman EM, Caldwell ME. Surface-plasmon spatial light modulators based on liquid crystal. Applied Optics. 1992; **31**(20):3880-3891. DOI: 10.1364/AO.31.003880

[29] Chiang KS, Liu WJ, Liao X, Rao YJ, Ran ZL. Laser-micromachined Fabry-Perot optical fiber tip sensor for high-resolution temperature-independent measurement of refractive index. Optics Express. 2008; **16**(3):2252-2263. DOI: 10.1364/OE.16.002252

[30] Allison G et al. A Fabry-Pérot cavity coupled surface plasmon photodiode for electrical biomolecular sensing. Nature Communications. 2021;**12**(1):1-7. DOI: 10.1038/s41467-021-26652-7

[31] Chavez G, Dinora A, Contreras C, Martin, Daniel JV. Sugar concentration sensor based on a micro Fabry-Perot interferometer. ECORFAN-Journal. 2017;**1**(2):16-21

[32] Martini J, Kiesel P, Roe JN, Bruce RH. Glucose concentration monitoring using a small Fabry-Pérot etalon. Journal of Biomedical Optics. 2009;**14**(3):034029. DOI: 10.1117/1.3153848

[33] Kumagai T, Tottori Y, Miyata R, Kajioka H. Glucose sensor with a Sagnac interference optical system. Applied Optics. 2014;**53**(4):720-726. DOI: 10.1364/AO.53.000720

[34] Winkler AM, Bonnema GT, Barton JK. Optical polarimetry for noninvasive glucose sensing enabled by Sagnac interferometry. Applied Optics. 2011;**50**(17):2719-2731. DOI: 10.1364/AO.50.002719

[35] Ann G, Li S, An Y, Wang H, Zhang X. Glucose sensor realized with photonic crystal fiber-based Sagnac interferometer. Optics Communication. 2017;**405**:143-146. DOI: 10.1016/J.OPTCOM.2017.08.003

[36] Chin LK, Lim CS, Liu AQ. A droplet Michelson interferometer for biochemical and protein analysis. Thirteenth International Conference on Miniaturized Systems for Chemistry and Life Sciences. Jeju. Korea. November 1–5. 2009. pp. 743-745

Chapter 6

Interferometric Gravitational Wave Detectors

Carlos Frajuca

Abstract

The existence of gravitational waves is an important proof of Einstein's theory of general relativity and took 100 years to be achieved using optical interferometry. This work describes how such a detector works and how it can change the way of seeing the Universe. Kilometers size laser interferometers are being built around the world in the way to make gravitational astronomy; detectors already built in the United States, Italy, and Japan will join efforts with detectors built in Japan and India and provide humanity with the means to see gravitational interactions of black holes and neutron star. Interactions, without these detectors, will be forever out of our sight.

Keywords: interferometric, gravitational waves, gravitational wave detectors, interferometric gravitational wave detectors, gravity

1. Introduction

Gravitational waves are ripples in the curvature of spacetime that propagate like waves, traveling outward from the source; they travel at the speed of light (299,792,458 m/s) and squeeze and stretch anything in their path as they pass. Do not confuse it with gravity waves that are waves generated when the force of gravity in a fluid medium or, when it is the case, at the interface between two different media, tries to restore equilibrium, as an example of these waves there are the wind waves on the interface between the atmosphere and the ocean.

Predicted in 1916 [1, 2] by Albert Einstein based on his theory of general relativity, [3] and detected in 2015, gravitational waves transport energy in the form of gravitational radiation, oscillation of spacetime itself. This theory predicts that the presence of mass causes spacetime to warp. When massive objects move around themselves, this curvature is altered, sending ripples of gravitation out of the universe carrying unbelievable amounts of energy. As these sources are very distant by the time, these disturbances catch up with us; they are almost imperceptible because they are weaker and gravitational waves interact very weakly with matter. Because of that, it was only a century after Einstein's prediction that scientists developed a sensitive enough detector—a Laser Interferometer Gravitational-Wave Detector, some kilometers long interferometer and were able to confirm the existence of gravitational waves [4].

The existence of gravitational waves is also a consequence of the Lorentz covariance of general relativity since it brings the concept of a finite speed of propagation of gravity.

Gravitational waves did not exist in the Newtonian theory of gravitation, which postulates that physical interactions propagate at infinite speed.

There was already indirect evidence of gravitational waves before its first direct detection. Measurements of the Hulse-Taylor binary system suggested that gravitational waves were more than a hypothetical concept. This system is one of the potential sources of detectable gravitational waves. These potential sources include binary compact star systems composed of white dwarfs, neutron stars, and black holes.

One way of thinking about gravitational radiation is as the messenger that carries information about changes in gravitational fields that attract one thing to another [5].

Several gravitational wave observatories (detectors) are under construction or in operation around the world [6]. In 2017, the Nobel Prize in Physics was awarded to Rainer Weiss, Kip Thorne, and Barry Barish for their role in detecting gravitational waves [7]. In these gravitational wave detectors and previous ones, the use of interferometry was essential for the operation of such a detector.

2. Some history

The possibility of gravitational waves was discussed in 1893 by Oliver Heaviside using the analogy between the inverse square law of distance in gravitation and electricity [7]. In 1905, Henri Poincaré proposed for the first time the existence of gravitational waves, which emanated from accelerated bodies and propagated at the same speed of light, this is dictated by the transformations of Lorentz [8] and implies an analogy that, accelerating electric charges produce electromagnetic waves, accelerating masses must emanate gravitational waves. When publishing his theory of gravitation (the general theory of relativity) in 1915, Einstein did not agree with Poincaré's proposal, as in his theory there are not gravitational dipoles, essential for the emission in the electromagnetism theory. However, based on a weak field approximation, he concluded that there should be three kinds of gravitational waves (named by Hermann Weyl as longitudinally-longitudinally, transversely-longitudinally, and transversely-transverse).

These approximations made by Einstein were criticized by several researchers and even Einstein had doubts. In 1922, Arthur Eddington wrote a paper entitled: "The propagation of gravitational waves" [8], in which he showed that two of the three types of waves proposed by Einstein were only mathematical artifacts produced by the system of coordinates and they were not really waves. This also cast doubt on the physicality of the third type (transversely transverse); however, Eddington proved that these would travel at the speed of light in all coordinate systems, so he did not rule out their existence.

In 1956, Felix Pirani corrected the confusion caused by the use of several coordinate systems by reformulating gravitational waves as the manifestation of the Riemann tensor observables. The Pirani work was ignored at that time mainly because the scientific community was concerned with another issue of whether gravitational waves could transport energy. This question was solved by Richard Feynman using a thought experiment presented at the first conference for General Relativity in 1957 known as the Chapel Hill Conference. His argument, known as the sticky bead argument, presents that: if a gravitational wave passes orthogonally to the beaded rod (a rod if some bead), the effect of it is to deform the bead and the rod, but as the rod is longer, the bead moves beads over the rod; this movement causes friction and then heat, which meant the passing gravitational wave would have energy. Afterward,

Hermann Bondi (who was skeptical of the existence of gravitational waves) published a more complete version of this argument.

After this conference, the scientific community took the existence of gravitational waves more seriously. Joseph Weber began to design and build a gravitational wave detector. It was the start of many gravitational wave detectors that are called Weber bars. Weber claimed to have detected gravitational waves in 1969 and 1970, the signals coming from the Galactic Center [9]. However, the high detection frequency quickly cast doubt on the validity of his observations, as the Milky Way's implied rate of energy loss would drain our galaxy's energy on a much shorter timescale than the galaxy's inferred age. It got worse when in the middle of the 1970 decade, the build of other Weber bar experiments by other groups around the world failed to detect such signs. By the end of the 1970 decade, the consensus was that Weber's detections were some kinds of noise.

The first indirect evidence of gravitational waves was discovered in 1974 by Russell Alan Hulse and Joseph Hooton Taylor Jr., using their discovery of the first binary pulsar. Results were published in 1979, showing the measure of the orbital period decay of the, so-called, Hulse-Taylor pulsar, which precisely describes the angular momentum and energy loss due to gravitational radiation emission predicted by general relativity. A discovery that gave them the 1993 Physics Nobel Prize.

This indirect detection of gravitational waves motivated further searches, despite Weber's discredited result. Some groups continued to improve on Weber's original concept. Using very low temperatures (cryogenic) for the bars, in high-vacuum systems and under vibrational isolation. There were many of these projects around the World. One of these groups built a Niobium bar resonant-mass gravitational wave detector [10]. In this detector, the vibrations caused by the passage of gravitational waves in the niobium bar are measured by a microwave parametric transducer. In this system, microwaves are pumped into a microwave cavity and the vibrations of the microwave cavity are connected to the niobium bar causing microwave signals in the microwaves to leave the cavity. This signal now must be amplified, but it mixes with the original microwave signal, which is too strong for the low-noise microwave amplifier; then the microwave carrier signal is removed by the use of an interferometer that cancels only the microwave carrier. A similar system can be seen in **Figure 1**. This is another use for interferometers in gravitational wave detectors; unfortunately, this kind of detector never made a detection, maybe because of a poor choice in the frequency range [11].

Other experimental groups pursued gravitational wave detection using laser interferometers. This idea of using appears to have been around for a long time by several independent groups, for example in 1962 ME Gertsenshtein and VI Pustovoit [12] and in 1966 by the group of Vladimir B. Braginskiĭ. The first prototype appeared in the 1970 decade built by Robert L. Forward and Rainer Weiss. In the following decades, increasingly sensitive detectors were constructed, culminating in LIGO and Virgo detectors.

After years and years of null results, the first detection of gravitational waves was made by LIGO on September 14, 2015, as the signal, named GW150914, probably came from the merger of two black holes [13, 14]. A year earlier, LIGO could have been brought down when scientists from the BICEP2 (Background Imaging of Cosmic Extragalactic Polarization 2) experiment claimed to have detected a weak signal in the CMB (Cosmic Microwave Background) that appeared like evidence of gravitational waves originating in the beginning universe. This evidence, according to researchers, could have been a smoking gun proof of the theory of cosmic inflation, which

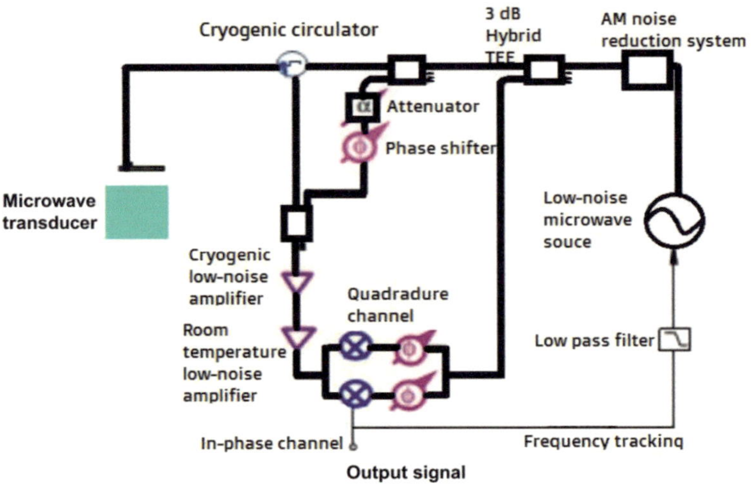

Figure 1.
Schematics of a similar electronics of Niobè gravitational wave detector showing a microwave interferometer (between the cryogenic circulator and the cryogenic low-noise amplifier) used to cancel the microwave carrier signal that will degrade the performance of the cryogenic low-noise amplifier.

postulates that very shortly after the Big Bang (10^{-32} seconds after), the expanding universe experienced a period of very rapid expansion (a factor of 10^{26} times). This fast expansion would have created ripples over the CMB, the fossil cosmic radiation that fills the universe being the first detectable electromagnetic radiation in Universe history. However, the BICEP2 signal detected could be explained also by Milky Way dust, making the scientists withdraw the claim that gravitational waves had been detected.

3. Detection

The interferometry system essentially works by measuring the variations that occur in light beams, which are arranged along two different arms. This analysis occurs when we observe the variations and interferences in the return of the light beams, which overlap, since according to the Theory of Relativity light always travels the same distance using the same time, this is our ideal ruler, eliminating the error of a form of measurement that also suffers from the geometric variations caused by ripples. All this technology must be sensitive enough to be able to detect variations of less than a thousandth of a proton.

A powerful laser beam passes through the beam splitter allowing the two generated beams to have the same phase and to be separated perpendicularly by the arms of 4 km each; at the end, they are reflected by the mirrors [13]. Everything was designed so that normally the phases of the waves of the originally emitted light beam and the reflected one generate a destructive effect, so nothing is detected by the photodetector. For the occasion of a gravitational wave passing by the Earth, causing spacetime expand and contract infinitesimally in one direction, thus generating interference arising from the physical property of the wave behavior of light when the phases produce a more constructive effect, thus a signal is detected. **Figure 2** shows the schematics of such a detector.

This is not so straightforward, as the gravitational wave passes through the detector, the gravitational wave also changes the spacetime between the two mirrors, if this

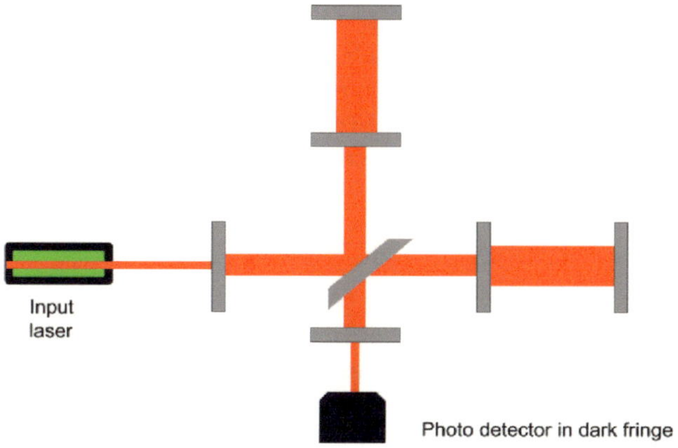

Figure 2.
Schematics of a laser-interferometric GW detector.

space was made of some material, the length of this material will also change by the gravitational wave, but the passage of gravitational wave does not change the speed of light, then the laser will travel in a shorter time in one arm and a larger time in the other arm then changing the interference pattern in the photodetector.

The fact that there are two observatories is a way of circumventing the possibility of confusing the detection of small earthquakes or some other local source of noise since when detecting a signal, this signal will be compared with that detected by the other observatory. It is only confirmed that this jolt was generated by gravitational waves if the generated signal has the same characteristics, for example, exactly the same profile in frequencies, since the observatories are exactly the same. Importantly, this all takes place in a vacuum, thus ensuring that the light will not have an unstable medium that could alter it in some way. Among the improvements to the observatories, the laser was updated to generate a higher frequency, there was an implementation of fused silica in the mirrors to reduce random mirror movements, and also their suspension was improved to reduce thermal noise and seismic isolation, making the observatories more sensitive for detection [14].

3.1 Indirect

In 1974, the most plausible indirect detection of gravitational waves was made by Joseph Taylor, Jr. and Russell Alan Hulse, when observing pulsars surrounding a neutron star (neutron stars are less dense cousins of black holes). That is the Hulse-Taylor binary, a pair of stars, which is a pulsar [15]. The characteristics of its orbit can be deduced from the Doppler shift of radio signals emitted by the pulsar. Each of the stars is about 1.4 M (M being the mass of the Sun) and the radius of their orbits is around 0.013 of the distance between Earth and Sun, smaller than the diameter of the Sun. The combination of smaller separation and bigger masses means that the energy emitted by the Hulse and Taylor binary will be much larger than the energy emitted by the Earth and Sun system by a factor of approximately 10^{22} times.

Information about the orbit can be used to estimate the amount of energy and angular momentum that would be emitted in the form of gravitational radiation. As the energy is emitted, the pulsars must get closer to each other. Measurements in the Hulse-Taylor system were performed over 30 years ago. The change in the orbital period corresponds to the prediction of gravitational radiation, assumed by general relativity to be within 1 part in 500. Russell Hulse and Joe Taylor, in 1993, were awarded the Physics Nobel Prize for this remarkable work, which was the first indirect evidence of gravitational radiation. This binary system, from the present until the time of the merger, is estimated to have a lifespan of a few hundred million years [16].

3.2 Direct

After that scientists were motivated to prove its existence. In the mid-1990s they were simulating the Black Hole merger, until the arrival of the Laser Interferometer Gravitational-Wave Observatory (LIGO) and the Italian Gravitational Wave Interferometer Detector (Virgo) at the end of the 20th century made slow progress needing much improvement. The challenge was not the physics itself or the equipment only, but the math behind it. Einstein's equations are called constraint equations, in which the solutions must always satisfy specific conditions, which is difficult since there are 10 equations with thousands of terms [17]. These simulations were necessary as the detectors needed a set of templates for the detected signal in order to establish the source of the signal, but using numerical relativity the set of these templates is still being built. Just to remember that Bayesian statistics is used in the detection.

In late 2015, researchers from the LIGO (Laser Interferometer Gravitational-Wave Observatory) project observed "distortions in spacetime" caused by a pair of black holes, about 30 solar masses each, in the process of merging [14, 17]. The signal was detected in the two LIGO sites on September 14, 2015 at 6:50:45 a.m. PDT, the event was named GW150914. The signal oscillated from 35 Hz to 250 Hz, with a time difference of 7×10^{-3} s between the detection of each observatory, and a maximum amplitude of 1.0×10^{-21}. Thus, coinciding with the shape predicted by Einstein almost exactly 100 years ago for an encounter of massive bodies, in the case of black holes that surround each other until they meet and merge, thus resulting in a significant warping in spacetime. To the relief of many, the merger took place at a distance of ~1.3 billion light-years from Earth. The masses of the initial black holes were 29 M and 36 M, (solar mass, M, approximately 1.99×10^{30} kg); the mass of the resulting black hole was 62M, and approximately 3.0 Mc^2 of energy was converted into gravitational waves to the rest of the Cosmos [14]. The signal of this measurement can be seen in Ref. [14]: the detected signal measured in the two observatories (located in Hanford, WA e Livingston, LA) are shown. First row shows the signal detected in the interferometers; it is shown a difference in time signal over the signal at the Louisiana observatory due to difference in the signal arrival time. Second row shows the expected signal of signal for the optimal source template previously calculated. Third row shows the residuals from rows 1 and 2. Row 4 shows the signal in the frequency domain against time.

In June 2016, the second burst of gravitational waves from merging black holes was announced, suggesting that such detections will soon become routine and part of a new kind of astronomy [18].

On June 1, 2017, for the third time, scientists announced that they had detected the infinitesimal reverberations of spacetime [19].

4. The real detectors

The interferometric gravitational wave detectors are very complex machines, besides being very big vacuum systems, need to have very powerful lasers, and so on.

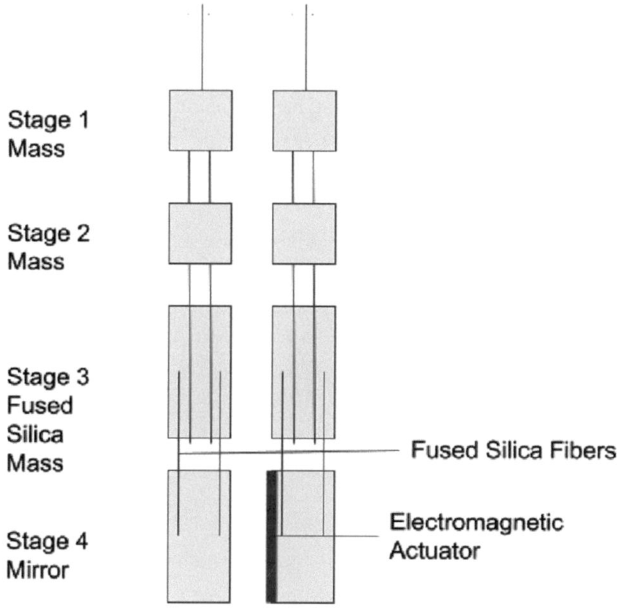

Figure 3.
Schematics side view of the mirror suspension system of the LIGO detector showing the electrostatic actuator which is used to keep the detector locked in.

The interferometer must be set in a dark fringe condition, but the mirrors are connected to the ground by the suspension, so they vibrate what could change the dark fringe condition. Then a very good suspension that attenuates the vibrations is used, **Figure 3** shows an example of such suspension (this example is about the LIGO detector). The Italian detector Virgo has a more sophisticated suspension which makes this detector more sensitive at lower frequencies.

The Virgo suspension is more complex, it is composed of an inverted pendulum and 6 masses suspended by its center, plus a collection of 18 LVDTs (Linear Variable Displacement Transducers), 5 accelerometers, 23 coils, three piezoelectric devices and 21 motor drivers. All these devices together are called the super attenuator. With a super attenuator for every suspended mirror.

These details show that it is very difficult to keep the interferometer locked in, it also depends on active systems.

But how sensitive the interferometer must be to make such measurements. The first measurement of gravitational waves was of displacement 10^{-18}, as the arms are 4 km long, and the variation in length was 4×10^{-15}. As the power inside the arms is 100 kW with an input power of 20 W, has a recycling factor of 5000 which makes the real sensitivity of the interferometer close to 10^{-12} m.

5. Conclusion

The Gravitational wave interferometric detector successfully detected gravitational waves with unprecedented sensibilities, for the position of the mirror the measurement precision was done in an order of 10^{-18} meters over square root of Hz. For such achievement, the mirror position should be stable for a factor of 10^{-13} m r.m.s.

Acknowledgements

The author acknowledges Conselho de Pesquisa e Desenvolvimento Cientifico (CNPq, Brazil) for Grant #312454/2021-0 and Fundacao de Amparo a Pesquisa do Estado de Sao paulo (FAPESP) Grant #2013/26258-4.

Conflict of interest

The authors declare no conflict of interest.

Author details

Carlos Frajuca
Rio Grande Federal University, Rio Grande, RS, Brazil

*Address all correspondence to: frajuca@gmail.com

References

[1] Einstein A. Näherungsweise Integration der Feldgleichugen der Gravitation. Sitzungsberichte der Königlich Preussischen Akademie der Wissenschaften Berlin. 1916; **part 1**:688-696

[2] Einstein A. Über Gravitationswellen. Sitzungsberichte der Königlich Preussischen Akademie der Wissenschaften Berlin. 1918; **part 1**:154-167

[3] Dave F. Einstein's gravity theory passes toughest test yet: Bizarre binary star system pushes study of relativity to new limits. Phys.Org. Internet. 2013. Available from: https://phys.org/news/2013-04-einstein-gravity-theory-toughest-bizarre.html

[4] What is a gravitational wave?: NASA Space Place. spaceplace.nasa.gov. Internet. 2017. Available from: https://spaceplace.nasa.gov/gravitational-waves/en/

[5] Collins H. Gravity's Shadow: The Search for Gravitational Waves. Chicago: University of Chicago Press; 2004

[6] Rainer Weiss, Barry Barish and Kip Thorne win 2017 Nobel Prize for Physics (in English). physicsworld.com. Internet. 2017. Available from: https://physicsworld.com/a/2017-nobel-prize-for-physics-winners/

[7] Heaviside O. A gravitational and electromagnetic analogy. The Electrician. 1893;**31**(18):5125-5134

[8] Stanley AE. The propagation of gravitational waves. Process Royal Society. 1922;**102**:268-282

[9] Levin J. Black Hole Blues and Other Songs from Outer Space. The Bodley Head Publishing House; 2016

[10] Tobar M, Locke CR, Ivanov EN, Heng IS, Blair DG. Accurate calibration technique for a resonant-mass gravitational detector. The Review of Scientific Instruments. 2000;**71**:4282-4285

[11] Frajuca C, Prado ARC, Souza MA. The challenge of calibrating a laser-inteferometric gravitational wave detector. Astronomische Nachrichten. 2021;**342**:115-122

[12] Gertsenshtein ME, Pustovoit VI. On the detection of low frequency gravitational waves. Soviet Physics JETP. 1963;**16**(2):433-435

[13] LIGO's interferometer. 2022. Available from: https://www.ligo.caltech.edu/page/ligos-ifo

[14] Abbott BP et al. Observation of gravitational waves from a binary black hole merger. Physical Review Letters. 2016;**116**(6):061102

[15] Weisberg JM, Taylor JH. Relativistic binary pulsar B1913+ 16: Thirty years of observations and analysis. 2004; arXiv:astro-ph/0407149

[16] Damour T. The discovery of the first binary pulsar. Classical and Quantum Gravity. 1974, 2015;**32**(12):124009

[17] Castelvecchi D, Witze W. Einstein's gravitational waves found at last. Nature News. 2016:19361

[18] Focus: LIGO Bags another Black Hole Merger. Journal "Physics". 2016;**9**:68

[19] LIGO snags another set of gravitational waves Spacetime vibrations arrive from black hole collision 3 billion light-years away by Emily Conover Internet. 2017. Available from: https://www.sciencenews.org/article/ligo-snags-another-set-gravitational-waves. [Accessed: February 15, 2022]